NEW DEVELOPMENTS IN ANXIETY DISORDERS

Edited by **Federico Durbano**
and **Barbara Marchesi**

New Developments in Anxiety Disorders
http://dx.doi.org/10.5772/62656
Edited by Federico Durbano and Barbara Marchesi

Contributors

Lu, Rudolf Gregurek, Alessandro Grecucci, Daniele Chiffi, Ferdinando Di Marzio, Remo Job, Jon Frederickson, Costina Păsărelu, Anca Dobrean, Janko Samardzic, Dubravka Svob Strac, Nesrin Dilbaz, Onat Yilmaz

Published by InTech

Janeza Trdine 9, 51000 Rijeka, Croatia

Publishing Process Manager Maja Bozicevic

Technical Editor SPi Global

Cover InTech Design team

First published December, 2016

Printed in Croatia

Additional hard copies can be obtained from orders@intechopen.com

New Developments in Anxiety Disorders, Edited by Federico Durbano and Barbara Marchesi

p. cm.

Print ISBN 978-953-51-2791-8
Online ISBN 978-953-51-2792-5

For EU product safety concerns:
IN TECH d.o.o., Prolaz Marije Krucifikse Kozulić 3, 51000 Rijeka, Croatia,
info@intechopen.com or visit our website at intechopen.com.

Meet the editors

Dr. Federico Durbano was born in Genoa, Italy, in 1963, and he is living near Milan, where he received his degree in Medicine and specialized in Psychiatry. He had different work experiences in some hospitals (Milan "Ospedale Maggiore Policlinico," Treviglio, Melegnano, Fatebenefratelli), where he has achieved significant career milestones, and now he is the head physician of Psychiatric Department in Melzo Hospital - ASST Melegnano e della Martesana. He had teaching assignments from the University of Milan (Nursing School) and University of Castellanza (Master in Criminology), he also attended more than 50 local and national congress and courses as invited speaker, and he published more than 150 papers. He also works as a technical advisor to the court in the field of forensic psychiatry.

Dr. Barbara Marchesi was born near Bergamo, Italy, in 1974, and she is living and working as a private practitioner in Milan. She received her degree in Psychology in Cattolica University of Milan; thereafter, she had a specialty course in Adolescent and Adult Psychodynamic Psychotherapy. After working for almost 10 years as a psychological consultant in Fatebenefratelli Hospital of Milan, where she completed also a Master in Psychodiagnostics, she worked 5 years in Papa Giovanni XXIII Hospital in Bergamo, where she completed the formation for Neuropsychology.

Contents

Preface

In this new volume, I have had the privilege to edit, there are some interesting updates on the clinical management of anxiety disorders.

The first three chapters addressed peculiar aspects of the neurobiology of anxiety, starting from different points of view than the classical neurobiological one.

In this respect, the authors of "Anxiety and Its Regulation: Neural Mechanisms and Regulation Techniques According to the Experiential-Dynamic Approach," adopting a peculiar psychological point of view, described an interesting model of emotional dysregulation (the Gross' process model) as an effective paradigm to explain why some psychological interventions are effective. The working hypothesis is based, according to the contributing authors, on the reinforcement of an appropriate regulatory strategy. The authors give some case descriptions to underline this theory, using the Experiential-Dynamic Emotion Regulation Model.

In the second chapter "Benzodiazepines and Anxiety Disorders: From Laboratory to Clinic," some data are given with regard to benzodiazepine treatment as a key element in understanding the mechanisms underlying the circuitry of anxiety itself. Also, some data on new pharmachological anxiety treatments, alternatives to benzodiazepines, are presented.

Last but not least, considering the neurological origins of Freud, a typical psychological paradigm such as psychodynamic could give interesting contributions to the understanding of the circuitry of anxiety, as explained in the chapter "The Integration of Psychodynamic Theories and Biological Aspects in the Development of Anxiety and Anxiety Disorders." The authors presented some data to describe the role of "visceral brain" neurons in the modulation of memory of fearful events and related somatic reactions. They also described an interesting case vignette in order to clarify their working hypothesis.

Two interesting and important contributions analyze the problem of the significant comorbidity of other anxiety disorders, not to mention other psychiatric disorders, specifically the peculiar relationship between anxiety disorders and substance abuse, both classical (alcohol as the most frequent) and those of more recent diffusion.

Anxiety, as such, is highly comorbid with a wide range of psychiatric disorders, suggesting a central role of emotion dysregulation in the clinical manifestations of psychiatric illnesses.

In the chapter "Comorbid Mental Disorders in Anxiety Disorders: Genetic Aspects of Bipolar Disorders and of Ethnicity," the authors examined the comorbidity issues of SUD and AD in a specific Chinese population compared to international data in order to underline the role of genetic factors.

In the next contribution, the authors of "Complex Comorbidity of Substance Use Disorders with Anxiety Disorders: Diagnosis and Treatment" analyzed the association of anxiety disorders and a specific diagnosis, substance use disorder (SUD), in term of epidemiology, treatment, and prognosis.

Albeit the data for the latter aspect are still temporary and insufficient, it seems important to open a debate on an issue that has become a kind of global pandemic, prompting a significant increase in the incidence of psychiatric disorders (especially acute cases in ER but also triggering early manifestations of other psychiatric problems that otherwise would have been either controlled or not fully expressed).

In relation to the rising importance social media platforms have on the development of social communication and therefore also on the expression of mental illness (in particular, social anxiety), there is also potential for more modern treatment approaches (see, e.g., the technological applications of telemedicine in the field of psychotherapeutic treatments). Accordingly, some authors who contributed to this volume described the important return of the involvement of family members in the treatment of anxiety disorders of adolescents and the role of modern social media in the expression of social anxiety.

The rapidly developing field of telemedicine, of which telepsychiatry is an equally rapidly growing aspect, is of particular interest in treating young patients deeply involved in digital communication platforms (Internet, social media groups, and so forth) and increasingly prone to developing an almost virtual relationship with the therapist. In this context, the authors of the chapter "Parental Involvement in Remotely Delivered CBT Interventions for Anxiety Problems in Children and Adolescents: A Systematic Review," starting from a CBT application for the treatment of anxiety, have involved parents in the therapeutic process with remotely delivered interventions for childhood anxiety disorders. They demonstrated the significant amelioration of clinical response of patients treated with the collaboration of parents. Their findings have allowed the development of a conceptual model to guide the future of research in the field.

Finally, in the very updated review "Impact of Social Media on Social Anxiety: A Systematic Review," the authors described the importance of the relationship of social media and new anxiety disorder manifestations, conducting a deep analysis of available and updated data. What is emerging is that there is a lack of information regarding the directionality of this relationship and its effect on mediators and moderators, suggesting the need of specific and sound researches in the field.

Therefore, this book offers readers some stimulating data about our future understanding and treatment of an "old" diagnosis as anxiety is.

Dr. Federico Durbano
Director of Psychiatric Unit "Martesana", Melzo, Italy
ASST Melegnano e della Martesana, Melzo Hospital, Milan, Italy
Contract Professor of Psychiatry at University of Milan Nursing School, Milan, Italy
Invited Professor to Criminology Master at "Il Sole 24 Ore" School of Economics, Milan, Italy

Dr. Barbara Marchesi
Consultant Psychologist, Psychotherapist, Expert in Psychodiagnosis and Neuropsychology
Milan, Italy

Chapter 1

Anxiety and its Regulation: Neural Mechanisms and Regulation Techniques According to the Experiential-Dynamic Approach

Alessandro Grecucci, Daniele Chiffi,
Ferdinando Di Marzio, Remo Job and
Jon Frederickson

Additional information is available at the end of the chapter

http://dx.doi.org/10.5772/65374

Abstract

Although anxiety is not necessarily a pathological phenomenon, it can become dysregulated, causing suffering. Indeed, emotion dysregulation lies at the core of many psychopathologies. Thus, anxiety regulation is central to all effective psychological treatment. The predominant perspective on emotion regulation and dysregulation is appraisal theory, which proposes that the cognitive appraisal of an event generates an emotional response. According to Gross's process model, any emotion can become dysregulated when the patient lacks or fails to use an appropriate regulatory strategy. Therefore, the clinician must teach the patient better regulatory strategies. The perspective we put forward departs from Gross's model based on appraisal theory. The experiential-dynamic emotion-regulation model, EDER, grounded in affective neuroscience and modern psychodynamic psychotherapy proposes that (1) emotions precede cognition (temporal and neuroanatomical primacy), (2) emotions are not inherently dysregulated (they have specific properties of time and strength proportional to the quality of the stimulus), and (3) dysregulation derives from the combination of emotions plus conditioned anxiety, or from secondary-defensive affects, both leading to dysregulated-affective states (DASs). To regulate DAS, the clinician must regulate the dysregulating anxiety or restructure the defenses, which create defensive affects, and then help the client to fully express the underlying emotions that elicit anxiety and defenses. In this chapter, we specifically focus on dysregulated anxiety, its neural bases, and how to regulate it according to the EDER model. First, we present hypotheses and data to show the neural bases of anxiety. Then, specific strategies and techniques to regulate anxiety are explained and clinical excerpts illustrate their application.

Keywords: anxiety, emotion regulation, affective neuroscience, psychotherapy

1. Introduction

We must study the mechanisms for regulating emotions when considering dysregulated emotions and emotionally driven behaviors in conditions such as personality or anxiety disorders [1–4]. The term emotion regulation refers to the neurocognitive mechanisms by which we regulate the onset, strength, and expression of our emotions [5]. According to Gross's process model of emotion regulation, emotions are generated through the following sequence: (1) an individual, exposed to a situation engages with it; (2) attends to a particular aspect of the situation; (3) interprets the event; (4) experiences an emotional response with a feeling, physiological arousal, and adaptive action tendency; and (5) the individual modulates that response. In this model, every emotion can in principle become dysregulated if the patient lacks or fails to use an appropriate regulatory strategy. Thus, the clinician must teach the patient better regulatory strategies. Cognitive-behavioral therapies (CBTs) follow these principles (see **Table 1**). Most of these strategies act at the level of attention and cognition and are consciously applied [4] by individuals to the experience of their emotions such as fear, anger, or sadness.

	Emotion generation and regulation in normality	**Emotion dysregulation in psychopathology**	**Implications for treatment**
Cognitive Emotion-Regulation model, CER (Gross [5] and further developments)	Cognitive appraisal produces emotions. The individual adopts cognitive top-down regulatory strategies at different levels (situation selection, situation modification, etc.) to regulate them	The individual fails to use regulatory strategies (deficit mechanism) and every emotion can become dysregulated	The clinician teaches the patient emotion-regulation strategies (therapeutic model: family of cognitive-behavioral therapies)
Experiential-Dynamic Emotion-Regulation model, EDER (Grecucci [3]; Frederickson and Grecucci [36]; Grecucci et al. [4])	Emotion is automatically generated by subcortical structures with certain properties (duration, intensity). The brain self-regulates emotions through a biological mechanism	After emotion is generated, dysregulatory mechanisms intervene that stop self-regulatory mechanisms and cause dysregulated-affective states (DASs)	The clinician helps the patient to remove dysregulatory mechanisms and downregulates DAS (therapeutic model: family of experiential- dynamic therapies)

Table 1. Two models of emotion regulation (modified from Grecucci et al. [4]).

Experiments studying emotion regulation show that individuals can learn to regulate their emotions, and their neural bases have been uncovered (see [6] for a review of basic findings). Of these, the dorsolateral prefrontal cortex, dlPFC, and the inferior parietal cortex, IPC, are commonly believed to control attention and working memory [7]. The anterior cingulate cortex, ACC, is associated with monitoring and control of ongoing processes [8]. The ventrolateral prefrontal cortex (v1PFC) appears to be responsible for selecting goal-appropriate responses [9, 10] and inhibiting inappropriate ones [11]. The target region of regulation is commonly believed to be the amygdala, a structure that supports the elaboration of external and internal emotional stimuli [12, 13] and negative stimuli [14].

In another line of research, Grecucci et al. [15–17] evaluated the regulation of socially cued emotions in real interactions. These studies showed that mentalizing (reappraising the intentions of the partner as less negative) changes emotional reactions, interpersonal behaviors, and neural responses. The tasks used in one of these experiments were derived from economic game theory (e.g., the ultimatum game and dictator game). One study showed that participants had weaker emotional reactions, used less rejection behavior, and had less neural activity if they downregulated their emotions while receiving unfair offers. This modulation of emotion was visible in an area of the brain involved in aversive reactions elicited by unfair offers, namely the insula. The insula has been found to represent visceral-affective experience [18–20], sensory experience [19], moral disgust, and anger [21].

A third line of research explored the interpersonal regulation of others' emotions. In line with [22, 23], we define this as regulation that occurs within interactions between two people. Despite the relevance of this type of regulation for clinical situations, we found only one attempt to study it in a laboratory setting. In a recent study [24], participants were asked to regulate their own (intrapersonal condition) and other persons' (interpersonal condition) emotional states. In the interpersonal condition, participants watched videos of people watching and reacting to the same emotional video they were watching. Participants were instructed to tell the person in the video how to interpret (reappraisal strategy) or suppress (suppression strategy) the emotional content of the video. Participants using interpersonal regulation showed decreased activation of the insula, the temporal-parietal junction, the temporal pole, and the medial prefrontal cortex similar to previous studies on emotion regulation of socially cued emotions [15, 17]. Despite methodological limitations, this study showed that regulating others' emotions interpersonally is possible.

1.1. Problems with cognitive regulation models

Cognitive theories of emotion regulation rely on the assumption that cognitive appraisals occur before emotional reactions, an assumption not supported by several neuroscientific studies of affect (see [25, 26], for a discussion). Emotion has a neurobiological primacy over cognition in terms of temporal dynamics (emotional stimuli are elaborated a few milliseconds before cognitive information [27]). The amygdala is activated and, in turn, activates the body before a later signal goes to the prefrontal cortex. Second, emotions have a neurobiological primacy over cognition in terms of anatomical circuitry, in that direct links exist between perceptual systems and emotional structures but not between perceptual systems and cognitive structures [25, 26].

Further, if cognition is primary, then cognitive strategies should be strong in the face of secondary emotional responses. However, research findings show that cognition-based strategies are not fully available for regulating emotions when emotion activation is high. For instance, experiments of emotion-regulation choice (how we choose which strategy to adopt in a given situation) [28] demonstrated that participants used reappraisal to regulate only low-intensity emotional stimuli and used distraction for high-intensity stimuli. This result reduces the importance of cognitive regulation strategies during stressful events that are more emotional than experimental stimuli used in the laboratory [26]. From a neurobiological point of view, a decrease in BOLD signal during induced emotional states in the prefrontal cortex (known to implement regulatory strategies) has been reported [29, 30].

2. Mechanisms of emotion generation, regulation, and dysregulation

Following the appraisal theory of emotion [31, 32], Gross' [5] emotion-regulation model holds that the cognitive appraisal of an event generates an emotional response. Based on this theory, cognitive-behavioral therapies focus on discrete cognitive, attentive, and behavioral factors to foster emotion regulation. In this view, emotion dysregulation occurs due to the failure to apply appropriate cognitive, attentive, and behavioral regulatory strategies (see [33–35]). Behavioral strategies (exposure to appropriate situations or adaptive modification of the situation), attentional strategies (increasing attentional flexibility or developing awareness to internal and external situational cues), and cognitive strategies (cognitive restructuring) are applied.

In this chapter, we depart from appraisal theory and cognitive emotion regulation, and present the experiential-dynamic model of emotion regulation [4, 36] that is grounded in affective neuroscience findings [25, 37, 38] and modern psychodynamic and experiential psychotherapy [39–43].

The experiential-dynamic emotion-regulation model [4] holds that events trigger (1) emotional responses prewired at birth [38] with inborn adaptive action tendencies [44] and facial expressions [45] which (2) precede cognition (temporal and neuroanatomical primacy) [46, 38]. According to neuroscientific research [25, 37, 38], emotions are automatically and implicitly generated by mainly subcortical brain structures with certain properties (duration and intensity proportional to the event) to give a sense of what is happening [46]. In normal conditions, the brain regulates emotions through a biological mechanism. They rise in intensity, peak, and then go flat with a Gaussian-like shape once the emotion's adaptive action tendency has been expressed in adaptive action.

Once elicited, emotions have a duration and intensity proportional to the stimulus and automatically self-regulate [26]. The conscious control or the use of a specific strategy is therefore not required to regulate emotions [4, 36]. Emotions are generated, expressed, and channeled into healthy actions and return to baseline [43] once the resulting adaptive action has served its evolutionary function. Thus, emotions are not inherently dysregulated [36].

Dysregulation derives from the combination of emotions plus conditioned anxiety, or from secondary-defensive affects, both leading to dysregulated-affective states (DASs). To regulate DAS, the clinician must regulate the dysregulating anxiety or restructure the defenses, which create defensive affects, and then help the client to fully express the underlying emotions that elicit anxiety and defenses (see **Table 1**).

But if emotions are not inherently dysregulated, what causes emotions to *become* dysregulated? Dysregulation results from (1) emotions paired with excessive levels of conditioned anxiety or (2) emotions that are triggered not by a discrete stimulus in reality but by an ongoing defense [43]. For instance, a patient projects that the therapist is criticizing him (imaginary and continual stimulus), and becomes afraid of the supposed criticism. Here, the patient is not afraid of the therapist; he is afraid of the image he places on the therapist (his projection). Thus, his ongoing anxiety results from his ongoing defense. Anxiety resulting from the defense of projection would be considered a defensive affect. We differentiate defensive affect (response to an imaginary stimulus) from true affect (response to a real stimulus).

The therapist following EDER model treats emotion dysregulation by applying experiential anxiety regulation techniques so that feelings can be explored without excessive anxiety or by restructuring the defenses that cause defensive affects. As a last step, the therapist helps the patient experience the underlying emotion fully without excessive anxiety or defenses so it can be channeled into adaptive action [39–43, 47]. Once the patient experiences his feelings deeply and channels them effectively into action, they no longer trigger anxiety, defenses, and symptoms instead (see [48]).

2.1. Why we regulate

Excessive anxiety causes painful physical symptoms due to activation of the somatic and autonomic nervous systems [49]. It compromises higher brain functions due to the release of neurohormones, which shut down the prefrontal cortex and hippocampus [43]. It triggers the use of defenses that create the patient's symptoms and presenting problems [33–35, 50, 51]. The compromised mental functioning and automatically triggered behavioral patterns prevent the patient from seeing better options, thinking about them, or being able to act on them. Excessive anxiety is a painful condition that Meltzer defines as unbearable mental pain that the mind wants to get rid of [52]. Thus, anxiety regulation is essential for any change to occur in psychotherapy.

To understand why and how anxiety is regulated in our brain and why "emotion regulation" does exist in the brain, we use an analogy from statistical mechanics (see [3] for a complete description). Statistical mechanics is the application of the theory of probability to the thermodynamic behavior of systems composed of a large number of particles. This branch of physics provides a model to link the microscopic properties of the individual elements to the macroscopic properties of the system made by them. Since the brain is one of the most complex systems in the universe (the brain is composed of 10^{12} interacting neurons), statistical mechanics can be fruitfully used to describe why and how mechanisms exist to regulate emotions. As the Boltzmann distribution suggests [53], the probability (P^1) that a system (the brain) converges toward a desired energetic state (E^1) (affective state) is negatively proportional to the temperature T^1 (level of dysregulated anxiety) of the system:

$$P^1 = Ke^{(-E1/T1)} \tag{1}$$

In every complex system, given two states E^1 and E^2, respectively, associated with temperatures T^1 and T^2, where T^2 is higher than T^1, the system will settle into state E^1, where the temperature is lower:

$$\frac{P^1}{P^2} = \frac{Ke^{(-E1/T1)}}{Ke^{(-E2/T2)}} \tag{2}$$

In other words, when emotion is activated, the mind can settle into that state (E^1 with its given T^1). However, if emotion is associated with excessive anxiety (DAS), the temperature overcomes the threshold of tolerability (say E^2 with T^2, where $T^2 > T^1$) and must be downregulated by the system. In our terms, for the brain to work, anxiety must be kept at acceptable levels (temperature T^1). The brain has several prewired mechanisms to self-regulate our emotions under normal conditions. However, if these mechanisms are interrupted by dysregulatory mechanisms, such as excessive anxiety (DAS), the temperature of the system will exceed the capacity of the mind to bear it. If this is the case, the system (in a state E^2 with T^2) will adopt strategies to abruptly lower it (defense mechanisms in psychodynamic terms that lower anxiety). However, the strategies to lower anxiety may have costs for the system (e.g., progressive cognitive, affective, behavioral distortions, and consequent symptom formation). One of the clinician's tasks is to help patients to regulate anxiety so they can experience progressively increasing levels of emotions without relying on defense mechanisms that create their presenting problems. After having examined **why** the brain needs regulatory mechanisms (excessive anxiety is painful), and **how** this is possible from a complex system perspective (the system constantly tries to move toward the lower energetic state), we focus on **what** is regulated.

2.2. What we regulate: anxiety

Emotions are evolutionary products designed to prime adaptive action [38, 54]. They become dysregulated by being paired with conditioned anxiety or they are created by dysregulating defenses [43]. If the therapist downregulates anxiety, the paired affect will become regulated. If the therapist restructures the defense that creates the dysregulated affect, the affect will disappear. For instance, if a patient can see the therapist accurately, the feelings toward the therapist based on the projection will disappear. After all, the defensive affects were an illusion based on the illusion of projection.

When an objective danger is detected, fear arises [50] to motivate us to deal adaptively with an objective threat. When a feeling arises, anxiety rises if it was dangerous to have this feeling in a previous relationship. Thus, anxiety is a signal that a feeling is rising that is potentially dangerous for this relationship [55]. All children in their development learn which feelings are allowed in their primary relationships and which feelings make caretakers anxious [56], thus threatening a relationship necessary for the child's survival [57–59]. To adapt [60], the child learns to ward off emotions that would threaten the relationship [56]. Thus, whenever a forbidden feeling arises, anxiety automatically signals danger [61]. The anxiety occurs out of the patient's awareness because it is generated nonconsciously in the brain [46].

The symptoms of anxiety in the body are created by the activation of the somatic and autonomic nervous systems. These systems are activated principally by several subcortical brain regions such as the cingulate and parahippocampal gyri, the amygdaloidal complex (the amygdala and bed nucleus of stria terminalis), septal nuclei, hypothalamus, some portions of the thalamus (inferior thalamic nuclei), and some parts of the basal ganglia [62]. Other brain areas involved in the anxiety circuit are the bed nucleus of the stria terminalis, hippocampus, hypothalamus, prefrontal cortex, periacqueductal gray matter, and the locus coeruleus.

The amygdala plays a fundamental role in the experience of anxiety by evaluating the valence (unpleasantness) and novelty of the stimulus [63–65]. We can divide the amygdala into three neural subgroups: the antero-central, the baso-lateral, and the medial. The baso-lateral group connects with the prefrontal orbital and medial cortex in the frontal lobe, and the associative cortex in the antero-temporal lobe. It is also hardwired with the bed nucleus of the stria terminalis, ventral hippocampus, and central amygdala. The antero-central group is hardwired with the hypothalamus and brain stem, including the parabrachial nucleus and the solitary tract nucleus [64–66]. Recent research shows that activating projection neurons in the baso-lateral amygdala (BLA) increases anxiety, while selective activation of axons from the BLA to the lateral central amygdala reduces anxiety. Activating the monosynaptic glutamatergic projection from the BLA to the ventral hippocampus also increases anxiety [67]. Sensory information enters the baso-lateral portion, and then relevant information goes to the central portion. The central amygdala modulates behavioral, physiological, and cognitive activity through its connection with the cortex and the hypothalamus. The bed nucleus of the stria terminalis, BNST, is responsible for linking anxiety to emotional stimuli. This area is connected to the amygdala, hypothalamus, and parabrachial nucleus. Indirectly, it is connected with the prefrontal cortex, hippocampus, hypothalamus, locus coeruleus, raphe nuclei, lateral septum, and periaqueductal gray matter. The BNST can be functionally divided into the anterodorsal (adBNST), ventral (vBNST), and oval (ovBNST) portions. Inhibiting the ovBNST, or exciting the adBNST reduces anxiety. By contrast, activating the ovBNST and inhibiting the adBNST cause anxiety [68]. Perhaps, in the absence of threatening stimuli, ovBNST is under the inhibitory control of the adBNST, or the adBNST is uninhibited by the inhibition of ovBNST. Inhibition of BLA neurons, connected to the adBNST, triggers anxious behavior. Conversely, their activation reduces anxiety as does selective activation of axons terminating in the lateral hypothalamus. The connection of the adBNST with the parabrachial nucleus mediates autonomic responses of anxiety [68]. vBNST has both glutamatergic excitatory synapses and GABAergic inhibitory ones with non-dopaminergic cells in the ventro-tegmental area (VTA) [68]. Glutamatergic inputs trigger avoidance and anxiety; the GABAergic inputs produce reward and reduce anxiety [69]. Glutamatergic vBNST cells show an increase of activity during a foot-shock session, while GABAergic cells are inhibited [70]. Septal nuclei in the anteromedial portion of forebrain receive inputs from the hippocampus and amygdaloidal complex and make synapses with the thalamus, hypothalamus, and brain stem. The septo-hippocampal system may mediate stress-induced anxiety [71].

2.3. Dysregulated anxiety

Emotions may be accompanied by excessive anxiety as a result of conditioning in previous attachments. Anxiety becomes a conditioned response indicating that a rising feeling could endanger a relationship [50, 58]. In turn, anxiety triggers defense mechanisms (cognitive distortion, behavioral avoidance) that cause patients' symptoms and presenting problems [72].

The combination of emotion and excessive anxiety creates a dysregulated-affect state [36]. An internal feeling is perceived as a threat in a relationship. The somatic and autonomic systems are activated [49, 54], creating symptoms of anxiety [49, 73]. This activation of the amygdala occurs before the message gets to the prefrontal cortex and becomes conscious [37].

The amygdala activates the somatic and autonomic nervous systems, which create a variety of anxiety symptoms in the body [42, 43]. When anxiety is discharged in the striated muscles (somatic nervous system), patients experience symptoms caused by tension in the striated muscles (tension in voluntary muscles, clenched hands, and sighing). When anxiety is discharged into the somatic nervous system, it is at a healthy level, so feelings can be expressed and experienced deeply [43].

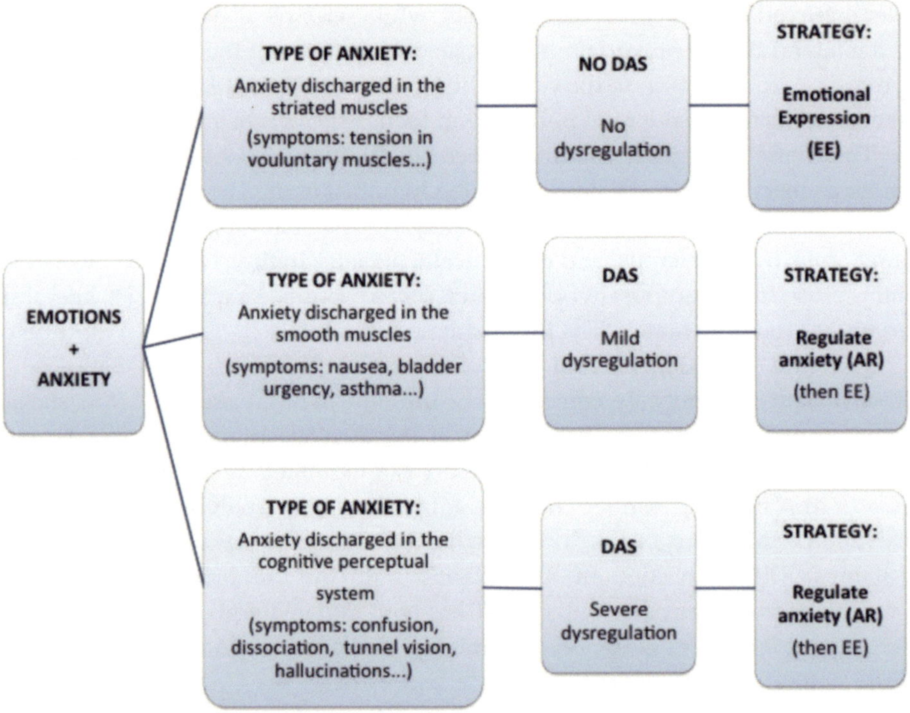

Figure 1. Decision tree for anxiety regulation.

When anxiety becomes too high, it shifts into the parasympathetic branch of the autonomic nervous system. Now, patients experience anxiety in the smooth muscles (nausea, diarrhea, migraines, sick to stomach, and need to use the bathroom), resulting in moderate DAS [74]. Hence, anxiety regulation becomes imperative [43].

When anxiety becomes even stronger, the anxiety is discharged even more into the parasympathetic nervous system, causing cognitive perceptual disruption, due to hypoperfusion of the prefrontal cortex (problems thinking, loss of reality testing, blurry vision, ringing in the ears, dizziness, fainting, and blanking out) [43]. The clinician should stop exploring emotions and regulate anxiety until it is discharged again into the striated muscles. See **Figure 1** for a summary of the proposed method (see [43] for a more detailed discussion).

When anxiety is too high, it triggers defenses, which can distort the perceptual/cognitive processes. For instance, denial can prevent a woman from seeing the danger of remaining with an abusive husband ("He didn't mean it.") Her defense prevents her from accurately seeing the stimulus to her feelings. Or she might deny that she has any negative feelings toward him for hitting her ("I love him."). Here, the defense prevents her from seeing her responses to his abuse, keeping emotions out of consciousness.

Likewise, defenses can prevent the patient from being aware of her anxiety and, thus, from regulating it. For instance, a patient might ignore her anxiety ("I always talk quickly. I'm just a fast talker."). Unable to see her anxiety, she cannot regulate it. We might hypothesize that defenses interfere with optimal communication between the prefrontal cortex and the amygdala. For anxiety regulation to occur, the patient must be able to observe her anxiety (the effects of the amygdala on the body). A recent study suggested that anxious behavior can occur because reduced functionality between the prefrontal cortex and the amygdala can disinhibit the amygdala [75].

3. Techniques to regulate anxiety

3.1. Problems with CBT techniques to regulate anxiety

Thinking can affect our emotions. But our emotional impulses can also channel and influence our thoughts [25, 46]. While modifying emotions gives rise to changes in cognitions, working primarily on changing cognitions may not necessarily cause a shift in emotion [25]. The CBT framework fails to understand that most feelings are not triggered by conscious thoughts but by nonconscious neural processes. In fact, most thoughts are defenses triggered by feelings. Cognitive interventions ask the patient to cognitively manage emotions based on willpower and repeated practice of a technique. The patient is being asked to consciously control anxiety and dysregulated-affect states while not being helped to address the cause, the unconscious feelings triggering the anxiety and defenses.

This problem results from the fact that CBT theory fails to recognize that core emotions (an evolutionary response to environmental and interpersonal cues) trigger anxiety, based on conditioned responses [55]. While anxiety can be regulated through attentive focus and self-soothing, it will continue to be triggered each time feelings arise. Unless the patient is helped to bear those previously warded-off feelings, the patient will not be helped to cope with the cause of his anxiety. In the following section, we outline regulatory strategies that are more consistent with affective neuroscience and the nonconscious processes that create emotions.

The CBTs are based on the assumption that the therapist can regulate the patient's anxiety. Unfortunately, this assumes that the patient is relating to the therapist instead of a projection placed upon the therapist. Just because the therapist wants to help does not mean the patient perceives her that way. In fact, if the patient perceives the therapist as critical, this perception will drive not only the patient's resistance to therapy, it will increase the patient's fear of the therapist. Anxiety cannot be regulated and the impasse cannot be resolved unless the therapist addresses the patient's projection.

Another potential drawback of the cognitive-behavioral framework is that techniques focusing on positive reappraisal may encourage some patients to obsessively ruminate about the thoughts that need reframing or pursue endless reframes without significant relief [76]. While the third-wave CBTs address this with cognitive defusion and mindfulness methods, this assumes that conscious top-down regulation can work with nonconscious bottom-up activation. See [4] for a more detailed discussion and a comparison between CBT and experiential-dynamic therapy (EDT) techniques.

3.2. Principles of experiential-dynamic techniques for regulating anxiety

Within the field of experiential-dynamic therapies, true feelings are understood to be generated through subcortical neuroperception of the environment, the experience of stimuli in reality [25, 37–39, 41–43, 46, 54, 55, 77]. DASs are created through the pairing of emotions with excessive conditioned anxiety or by defenses which create defensive affects [39, 41–43]. With this dual theory of causation, we differentiate true feelings generated by real stimuli from defensive affects generated by imaginary stimuli.

In the course of therapy, the patient describes relationships where his problem occurs. The therapist explores the feelings triggered toward a specific person for doing a specific thing to the patient. As feelings rise, feelings trigger anxiety and defenses. If necessary, the therapist regulates anxiety and then explores the feeling. Otherwise, the therapist helps the patient see the defenses that create his symptoms and presenting problems and then encourages the patient to face his feelings rather than use his defenses.

Sometimes, anxiety can be regulated fairly easily by helping the patient pay attention to the physical experience of anxiety in the body. Or the patient may be helped to feel less anxious just by understanding the process of the session: as he explores his feelings, feelings trigger anxiety, anxiety triggers defenses, and the defenses create his symptoms and presenting problems. Understanding causality orients the patient, reducing anxiety.

Another way to regulate anxiety is to block defenses that perpetuate or escalate anxiety. For instance, rumination about past events or future fantasies will perpetuate anxiety. Blocking those defenses will block the rise of anxiety based on fantasies. Then, the therapist directs the patient's attention to the feeling in this moment that triggers anxiety in this moment. Keeping the focus in the here and now maintains an effective focus that will be inherently anxiety regulating. The therapist needs to ask himself: (1) is the patient focusing on experience in the past, future, or present? (2) Is this a real stimulus or an imaginary one? (3) Is this a specific example or a vague one? And (4) what is the patient feeling toward whom for doing what? To explore feelings and regulate anxiety, we need to examine what the patient feels now regarding a real stimulus toward a real person for a real deed. This clarity and focus is inherently regulating.

Every time anxiety moves into the smooth muscles or cognitive/perceptual disruption, the combination of feeling and anxiety creates a DAS. The therapist stops exploring feeling and regulates anxiety until it returns to the striated muscles. The feeling, without dysregulating anxiety, is inherently regulated. If the therapist fails to regulate the anxiety until it is in the striated muscles, the patient will suffer from more severe somatic symptoms and will shift into more primitive defenses that lead to a loss of reality testing, which will perpetuate his anxiety. While assessing the patient's anxiety, the therapist will monitor how rapidly it rises, how slowly it drops as a result of regulation, and the pathway of anxiety discharge in the patient's body (somatic or autonomic nervous system). This assessment allows the therapist to ensure that anxiety has been regulated enough that it is safe to explore feelings to build the patient's capacity for affect tolerance.

Every story the patient tells triggers feeling and anxiety, allowing the therapist to assess what issues and feelings trigger the most feelings, anxiety, and defenses. Obviously, whatever issue triggers the most anxiety is where the patient needs the most help. Each time anxiety rises, the therapist helps the patient see the issue he has, the anxiety it triggers in his body, the defenses he uses, and how those defenses create the patient's symptoms and presenting problems. If the patient can observe his feelings, his anxiety, his defenses, and how those defenses create his presenting problems, he can understand what causes his suffering and the therapeutic task: let go of the defenses causing his suffering and face the feelings he has been avoiding.

When anxiety moves out of the striated muscles into the smooth muscles or cognitive/perceptual disruption, the therapist should regulate the anxiety and then explore feeling gradually, what we call the "graded format" [39, 40, 43]. In the graded format, the therapist explores feelings gradually until anxiety moves into the smooth muscles or cognitive/perceptual disruption. Then, he stops exploring feelings and regulates anxiety until it returns to the striated muscles. Then, the therapist explores feelings again. Each time the therapist and patient explore feelings at progressively higher levels until the patient can experience the full extent of his feelings without his anxiety moving out of the striated muscles and without using defenses that cause DAS.

We have described earlier how affects by themselves are not dysregulated. They are adaptive responses that guide mammalian and human behavior, proportional to the stimulus. When affects are dysregulated by excessive anxiety, we regulate anxiety and, thereby, end the affect dysregulation. However, with mention of defenses that cause DAS, we shift to a second source of dysregulated affects: affects that are triggered by defenses, what we call defensive affects.

Simply put, if a person criticizes you unfairly (real stimulus), you will feel anger toward that person. However, if you *imagine* a person is critical and wants to hurt you (imaginary stimulus), you will feel angry also. You would not be angry at what that person really did but at the projection you place on that person. Now your feeling results from a defense: projection. That is why we call it a defensive affect. It is the result of an imaginary stimulus, not a real stimulus.

Likewise, you might become afraid, fearing at any moment that this imaginary critic will attack you. This is not the fear of an objective danger, this is anxiety triggered by a projection. That is why we call it projective anxiety [43, 78]. So, if you project that someone wants to criticize you, you might become angry with that person or afraid of that person. In either case, the anger or anxiety would be defensive affects, feelings that result from the defense of projection.

Let us clarify causality more deeply. A patient is angry with his boss. This anger triggers anxiety. But he denies that he is angry and projects that his boss is angry with him. In response to this projection, he becomes afraid of a supposedly angry boss. Now, anxiety regulation will do no good. We can regulate anxiety that is caused by a genuine feeling toward a real person for a real deed. But we cannot regulate anxiety that is caused by an *ongoing* projection. For as long as he projects onto the boss, he will be afraid of the boss, or, more precisely, the projection he places on the boss. First, we must help the patient see how he is projecting onto his boss.

What impact can this insight have on therapy? To deal with the anxiety of having feelings or desires in therapy, the patient may project those feelings or desires onto the therapist. For instance, an angry patient may project that the therapist is angry. The patient who wants to look at her inner life may project that the therapist wants to look at her inner life. In each case, the patient becomes afraid of the therapist's imagined feeling or desire. This is projective anxiety: fear of the projection.

The therapist cannot regulate a patient's anxiety if the patient is projecting onto the therapist. If the patient projects that the therapist is critical, a critic cannot regulate anxiety. The therapist must deactivate the projection that is perpetuating the patient's high anxiety. Once the patient sees the therapist realistically, the anxiety due to projection will disappear. Then the therapist can explore the patient's inner feelings and desires she formerly projected onto the therapist.

The reader may be puzzled since it is well known that defenses are unconscious mechanisms for avoiding forbidden feelings and the anxiety they trigger [50, 51]. Defenses are supposed to *reduce* anxiety. That is true for some defenses. The EDER therapist focuses on the avoided feelings that trigger the patient's anxiety. If anxiety rises, the therapist knows he is approaching the issues the patient avoids. Thus, anxiety is a good sign: the therapist is approaching important therapeutic material. If anxiety becomes too high, it must be regulated. If anxiety remains in the striated muscles, it does not need to be regulated.

When feelings trigger anxiety, patients use defenses to ward off feelings and the anxiety. For instance, when the patient uses a defense, such as intellectualization, his awareness of feeling drops and his anxiety will drop. If the therapist identifies the defense for the patient and invites him to face the feelings underneath his intellectualization, the patient's anxiety will rise because the patient and therapist are going toward the avoided feelings that trigger anxiety.

However, certain defenses do not reduce anxiety. For example, the patient may project his anger onto the therapist to avoid feeling it within himself. However, once he projects his anger onto the therapist, he can become afraid of the therapist, resulting in even more anxiety, creating a DAS.

Suppose a man assaults a woman (stimulus in reality). This triggers anger (true feeling) in her and, as a result, she is able to fight him off. However, in the therapist's office she is terrified of the therapist, imagining that he is angry. She projects her anger upon the therapist (defense). This defense of projection creates fear (defensive affect), the result of projecting upon the therapist. Or suppose this patient criticizes herself for how she handled the assault. She turns the anger toward the assailant onto herself (defense) and becomes sad (defensive affect).

Understanding what causes a given feeling allows us to intervene effectively. If the patient's fear results from projection, we need to deactivate the projection [42, 43, 79] so that the anxiety resulting from projection will drop. Here, cognitive and experiential-dynamic therapists agree. Likewise, if the patient is sad due to self-attack, we need to help the patient see the defense and relinquish it, so that her defense-caused sadness (defensive affect) will drop. Again, cognitive and experiential-dynamic therapists agree.

Clinically, one form of dysregulated emotion is true feeling plus excessive anxiety paired through conditioning. Classic anxiety regulation techniques shared by cognitive and experiential-dynamic therapists bring the patient's anxiety down until the patient can bear her underlying feeling without anxiety [39, 43, 79]. However, to prevent future relapse, a risk in cognitive therapies (see [80, 81]), once the defense has been relinquished, experiential-dynamic therapists will explore the true feeling underneath, which triggered the anxiety and defenses [39, 41–43].

Unless the therapist builds the amount of feeling the patient can bear, DAS will occur with each subsequent activation of feeling (see [82] for a review illustrating the relationship between the degree of emotional experience and the level of long-term outcome.) To prevent further DAS in the future, the experiential-dynamic therapist will explore the feeling at progressively higher levels. Each time anxiety gets too high, the therapist will regulate anxiety (downregulation), and then explore the feeling at a higher level.

In this gradual stepwise exposure method, the therapist builds the patient's capacity to bear the full extent of her feelings without becoming dysregulated by anxiety or defensive affects. In this model, excessive anxiety is determined by whether the patient's anxiety shifts into the parasympathetic branch of the autonomic nervous system (see [43] for a fuller discussion of the symptoms, which indicate that the patient has gone over the threshold of anxiety tolerance, and signs of cognitive impairment due to neurohormonal discharge).

This graded exposure to feelings helps the patient develop the capacity to bear her feelings without anxiety, so she can channel them into effective action [40, 43]. Once she can bear her feeling to the fullest extent, relapse into DAS can be prevented.

A second form of emotion dysregulation occurs when the patient's defenses cause a defensive affect [39–43]. For instance, a patient who is irritated with the therapist may use the defense of self-attack and become depressed in session. The therapist will help the patient face the feeling toward him without using the defense of self-attack, which is causing the DAS. "If we look under these critical thoughts, I wonder what feelings might be coming up here with me?" The therapist helps build the patient's capacity to face and label her feelings without using the defense of self-attack creating her DAS. See [36] for a more detailed discussion on how to deal with DAS due to defensive affects.

Type of affect	Strategy	Technique
Emotions	Emotional expression	-Identify and label the emotions -Help the patient pay attention to the emotions in the body -Differentiate feelings from anxiety -Differentiate feelings from defense -Experience feelings physically in the body -Feel the impulse physically in the body -Portray the impulse
DAS due to excessive anxiety	Regulation of anxiety	-Identify the symptoms of anxiety in the body -Mobilize self-observing capacity -Pay attention to anxiety in the present moment -Differentiate anxiety from the stimulus that generated it -Show causality: feelings in this moment trigger anxiety that triggers symptoms -Differentiate the symptoms of anxiety from the experience of feelings -Interrupt defenses that prevent anxiety regulation (ignoring, avoiding the present moment) - Interrupt defenses that perpetuate anxiety (self-attack, projective anxiety, symbolic equation) - Address spatial and temporal distortions -Shift the resistance system to isolation of affect -Restructure the pathway of anxiety discharge from the smooth muscles or cognitive/perceptual disruption into the striated muscles (the graded format) -Interrupt projective anxiety

Table 2. Strategies and techniques to regulate anxiety in the EDER model.

When the patient responds with feeling, the therapist will encourage the patient to experience her feeling more deeply, "How do you experience that anger physically in your body?" Or if the patient becomes sick to her stomach (a sign of anxiety), the therapist will intervene immediately. "That's a sign of anxiety. If we look under the anxiety, could we take a look under your anxiety and see what feelings are coming up here toward me?" If the patient becomes irritated toward the therapist and then shifts into a DAS of weepiness, the therapist will interrupt the defense immediately causing her DAS. "Notice how these tears come in to wash away your anger? Could they be making you depressed? Could they be protecting me? If you don't protect me, could we look underneath those tears and see how you experience the anger that's underneath those tears?"

In these examples, we see how different defenses create different defensive affects. We either restructure the defense (e.g., projection) to eliminate the anxiety or defensive affects resulting from projection, or we identify the defense (e.g., self-attack) and DAS and then help the patient face the feelings which the defense and DAS are covering. In these ways, we build the patient's capacity to identify, experience, and bear her feelings without anxiety or defenses. Then, she can channel those feelings into effective action so they can fulfill their original evolutionary purpose: adaptation (see **Table 2**).

4. Conclusion

The EDER model presented in this chapter departs from other emotion-regulating psychotherapies (see, e.g., [34, 35, 83, 84]). One of the assumptions is that emotion activation is a nonconscious physiological process that occurs initially without conscious awareness. The brain is programmed to generate emotions in response to internal and external stimuli. Emotions have physical properties with intensity and duration proportional to the intensity of the stimulus. Emotions, as evolutionary products, are not inherently dysregulated. They are activated so they can be channeled into adaptive action.

Moreover, emotion dysregulation is conceptualized as the result of dysregulating anxiety and defenses, not by the lack or failure of regulatory strategies, thus departing from a CBT view. From this perspective, dysregulation results from (1) emotions paired with excessive conditioned anxiety in the parasympathetic branch of the autonomic nervous system and (2) affects created not by a stimulus in reality but by defenses such as projection and self-attack. Frederickson and Grecucci [48] define feeling plus excessive anxiety and feelings covered by defensive affects as dysregulated-affective states. DAS due to excessive anxiety needs to be regulated [3, 4, 48]. If anxiety is too high, it impairs reality testing, and cognition is impaired due to a shutdown of the prefrontal cortex. Once anxiety is regulated and returns to the striated muscles and reality testing is restored, cognitive reworking (in a psychodynamic or CBT fashion) can be done. We use two basic overarching strategies: (1) emotion expression for true feelings triggered by real stimuli and (2) regulation of anxiety (and restructuring of defenses) for DAS due to emotions paired with excessive anxiety (see [43], for more details). We then explore true feelings as deeply as the patient can bear. The therapist upregulates true emotions while deactivating DAS until the patient fully experiences the previously avoided emotions without the DAS caused by anxiety (or by defenses). Now, the patient feels relief and can channel emotions into effective, adaptive action.

Author details

Alessandro Grecucci[1*], Daniele Chiffi[2], Ferdinando Di Marzio[1], Remo Job[1] and Jon Frederickson[3]

*Address all correspondence to: alessandro.grecucci@unitn.it

1 Department of Psychology and Cognitive Sciences, University of Trento, Trento, Italy

2 Department of Molecular Medicine, University of Padova, Padova, Italy

3 Washington School of Psychiatry, Washington, DC, USA

References

[1] Kring, A.M., Werner, K.H. Emotion regulation and psychopathology. In: P. Philippot and R.S. Feldman, editor. The regulation of emotion. Hove, UK: Psychology Press; 2004. p. 359–385.

[2] Ochsner, KN1, Gross JJ2. Cognitive Emotion Regulation: Insights from Social Cognitive and Affective Neuroscience. Curr Dir Psychol Sci. 2008 Apr;17(2):153–158. DOI: 10.1111/j.1467-8721.2008.00566.x.

[3] Grecucci, A. The Epistemologic Conflict. Psychoanalysis and Neuroscience of anti-knowledge processes. (Il conflitto epistemologico. Psicoanalisi e Neuroscienze dei processi anticonoscitivi). Francavilla al Mare: Edizioni Psiconline; 2012.

[4] Grecucci, A., Theuninck, A., Frederickson, J., Job, R. Mechanisms of social emotion regulation: from neuroscience to psychotherapy. In: Madeline L. Bryant, editor. Emotion Regulation: Processes, Cognitive Effects and Social Consequences. Nova Science Publishers; 2015. p. 57–84.

[5] Gross, J.J. The emerging field of emotion regulation: an integrative review. Review of General Psychology. 1998;2(3):271–299. DOI: 10.1037/1089-2680.2.3.271.

[6] Gross, J.J., Ochsner, K.M. The cognitive control of emotion. Trends in Cognitive Sciences. 2005;9(5):242–249. DOI: 10.1016/j.tics.2005.03.010.

[7] Corbetta, M., Shulman, G.L. Control of goal-directed and stimulus-driven attention in the brain. Nature Reviews Neuroscience. 2002; DOI: 10.1038/nrn755.

[8] Paus, T. Primate anterior cingulate cortex: where motor control, drive and cognition interface. Nature Reviews Neuroscience. 2001;2(6):417–424. DOI: 10.1038/35077500.

[9] Badre, D. Wagner, AD. Left ventrolateral prefrontal cortex and the cognitive control of memory. 2007; DOI: 10.1016/j.neuropsychologia.2007.06.015.

[10] Novick, J.M., Trueswell, J.C., Thompson-Schill, S.L. Cognitive control and parsing: reexamining the role of Broca's area in sentence comprehension. Cognitive, Affective, & Behavioral Neuroscience. 2005;5(3):263–281.

[11] Aron AR, Robbins TW, Poldrack RA. Inhibition and the right inferior frontal cortex. 2004; Trends Cogn Sci. 2014 Apr; 18(4):177–85. DOI: 10.1016/j.tics.2013.12.003. Epub 2014 Jan 15.

[12] Van Bavel, J.J., Packer, D.J., Cunningham, W.A. The neural substrates of in-group bias: a functional magnetic resonance imaging investigation. Psychological Science. 2008;19:1131–1139.

[13] Van Bavel, J.J., Cunningham, W.A. A social neuroscience approach to self and social. European Review of Social Psychology. 2011;21:237–284.

[14] Whalen, P.J., Kagan J., Cook R.G., et al. Human amygdala responsivity to masked fearful eye whites. Science. 2004;306(5704):2061. DOI: 10.1126/science.1103617.

[15] Grecucci, A., Giorgetta, C., van Wout, M., Bonini, N., Sanfey, A.G. Reappraising the ultimatum: an fMRI study of emotion regulation and decision-making. Cerebral Cortex. 2013a;23(2):399–410. DOI: 10.1093/cercor/bhs028.

[16] Grecucci, A., Giorgetta, C., Bonini, N., Sanfey, A.G. Living emotions, avoiding emotions: behavioral investigation of the regulation of socially driven emotions. Frontiers in Psychology. 2013b;3:616.

[17] Grecucci, A., Giorgetta, C., Bonini, N., Sanfey, A.G. Reappraising social emotions: the role of inferior frontal gyrus, temporo-parietal junction and insula in interpersonal regulation. Frontiers in Human Neuroscience. 2013c;7:523.

[18] Mufson, E.J., Mesulam, M.M. Insula of the old world monkey. Journal of Comparative Neurology. 1982;212(1):1–52. DOI: 10.1002/cne.902120102 - 10.1002/cne.902120103 - 10.1002/cne.902120104.

[19] Craig, A.D. How do you feel—now? The anterior insula and human awareness. Nature Reviews Neuroscience. 2009; DOI: 10.1038/nrn2555.

[20] Wager, T.D., Barrett, L.F. (2004). From affect to control: Functional specialization of the insula in motivation and regulation. Published online at PsycExtra.

[21] Sanfey, A.G., Rilling, J.K., Aronson, J.A., Nystrom, L.E., Cohen, J.D. The neural basis of economic decision-making in the ultimatum game. Science. 2003;300:1755–1758.

[22] Niven, K., Totterdell, P., Holman, D. A classification of controlled interpersonal affect regulation strategies. Emotion. 2009;9(4):498–509. DOI: 10.1037/a0015962.

[23] Zaki, J., Williams, W.C. Interpersonal emotion regulation. Emotion. 2013;13(5):803–810. DOI: 10.1037/a0033839.

[24] Hallam, G.P., Webb, T.L., Sheeran, P., Miles, E., Niven, K., Wilkinson, I.D., Hunter, M.D., Woodruff, P.W.R., Totterdell, P., Farrow, T.F.D. The neural correlates of regulating another person's emotions: an exploratory fMRI study. Frontiers in Human Neuroscience. 2014;8:376. DOI: 10.3389/fnhum.2014.00376.

[25] Panksepp, J., Biven, L. The Archaeology of Mind: Neuroevolutionary Origins of Human Emotions. New York, NY: W.W. Norton; 2012.

[26] Grecucci, A., Job, R. Rethinking reappraisal: insights from affective neuroscience. Behavioral and Brain Sciences. 2015;38:30–31. DOI: 10.1017/S0140525X14001538.

[27] LeDoux, J. The Emotional Brain. London: Weidenfeld & Nicolson; 1998.

[28] Sheppes, G., Scheibe, S., Suri, G., Gross, J.J. Emotion-regulation choice. Psychological Science. 2011;22(11):1391–1396. DOI: 10.1177/0956797611418350.

[29] Mayberg, H.S., Liotti, M., Brannan, S.K., McGinnis, S., Mahurin, R.K., Jerabek, P.A., Silva, J.A., Tekell, J.L., Martin, C.C., Lancaster, J.L., Fox, P.T. Reciprocal limbic-cortical function and negative mood: converging PET findings in depression and normal sadness. The American Journal of Psychiatry. 1999;156(5):675–682. DOI: 10.1176/ajp.156.5.675.

[30] Dolcos, F., McCarthy, G. Brain systems mediating cognitive interference by emotional distraction. The Journal of Neuroscience. 2006;26(10). DOI: 10.1523/JNEUROSCI.5042-05.2006.

[31] Frijda, N., Mesquita, B. The analysis of emotions: dimensions of variation. In: Mascolo, M.F., Griffin, S., editors. What Develops in Emotional Development? New York: Plenum; 1998: p. 273–295. DOI: 10.1007/978-1-4899-1939-7_11.

[32] Scherer, K.R. Appraisal considered as a process of multi-level sequential checking. In: Scherer, K.R., Schorr, A., Johnstone, T., editors. Appraisal Processes in Emotion: Theory, Methods, Research. New York, NY, and Oxford: Oxford University Press; 2001. p. 92–120.

[33] Beck, A,T. (1975). *Cognitive therapy and the emotional disorders*. Madison, CT: International Universities Press, Inc. ISBN 0-8236-0990-1

[34] Linehan, M.M. Cognitive-behavioral treatment of borderline personality disorder. New York, NY: Guilford. Press; 1993a.

[35] Linehan, M.M. Skills training manual for treating borderline personality disorder. New York, NY: Guilford Press; 1993b.

[36] Frederickson, J., Grecucci, A. (under review). Dysregulated Affects and dysregulating Defenses: toward an emotion regulaiton informed dynamic psychotherapy.

[37] LeDoux, 1996. The Emotional Brain. London: Weidenfeld and Nicolson.

[38] Panksepp, J. The Origins of Affective Neuroscience: The Foundations of Human and Animal Emotions. Oxford: Oxford University Press; 1998.

[39] Coughlin della Selva, P. Intensive Short-Term Dynamic Psychotherapy. New York, NY: Karnac Books; 1996.

[40] Allan, A. Reaching through resistance. Advanced Psychotherapy Techniques. Kansas City, MO: Seven Leaves Press; 2015.

[41] Davanloo, H. Unlocking the Unconscious. Chichester, England: John Wiley & Sons; 1990. p. 283–306.

[42] Davanloo, H. Intensive short-term dynamic psychotherapy—central dynamic sequence: phase of pressure. In: Davanloo, H., editor. Intensive Short-Term Dynamic Psychotherapy: Selected Papers of Habib Davanloo. Chichester, England: John Wiley & Sons; 2000. p. 183–208.

[43] Frederickson, J. Co-Creating Change: Effective dynamic therapy techniques. Kansas City, Mo: Seven Leaves Press; 2013.

[44] Frijda, N. The Emotions. Cambridge: Cambridge University Press; 1986.

[45] Ekman, P. Emotions Revealed: Recognizing Faces and Feelings to Improve Communication and Emotional Life. New York, NY: Oxford University Press; 2003.

[46] Damasio, A. The Feeling of What Happens: Body and Emotion in the Making of Consciousness. New York, NY: Harcourt Press; 1999. 386 p.

[47] McCullough, L., Kuhn, N., Andrews, S., Kaplan, A., Wolf, J., Hurley, C. Treating Affect Phobia: A Manual for Short-Term Dynamic Psychotherapy. Guilford Press; 2003.

[48] Frederickson, J., Grecucci, A. (under revision). Dysregulated Affects and Dysregulating Defenses: toward an emotion regulation informed dynamic psychotherapy. Frontiers in Psychology.

[49] Ochsner, K.N., Ray, R.D., Cooper, J.C., Robertson, E.R., Chopra, S., Gabrieli, J.D., Gross, J.J. For better or for worse: neural systems supporting the cognitive down- and upregulation of negative emotion. NeuroImage. 2004;23(2):483–499. DOI: 10.1016/j.neuroimage.2004.06.030.

[50] Freud, A. The Ego and the Mechanisms of Defense. Madison, Conn: International Universities Press; 1936.

[51] Freud, S. The Ego and the Id. 1923. (A. Strachey, Trans. And J. Strachey, Ed.). New York: NY. (Original work published 1923).

[52] Meltzer, 1986. Studies in Extended Metapsychology: Clinical Applications of Bion's ideas (Strath Tay, Pertshire: Clunie Press).

[53] Ackley, D.H., Hinton, G.E., Sejonowski, T.J. A learning algorithm for Boltzmann machines. Cognitive Science. 1985;9:147–169.

[54] Porges, S. The Polyvagal Theory: Neurophysiological Foundations of Emotions, Attachment, Communication, and Self-Regulation. New York, NY: W.W. Norton; 2011.

[55] Freud, S. Inhibitions, Symptoms, Anxiety. In: Strachey, A. (Trans), Strachey, J., editors. New York, NY: Norton; 1926.

[56] Sullivan, H.S. Interpersonal Theory of Psychiatry. 1953. New York, NY: W.W. Norton.

[57] Bowlby, J. Attachment and Loss, Volume One: Attachment. New York, NY: Basic Books; 1969.

[58] Bowlby, J. Attachment and Loss, Volume Two: Separation-Anxiety and Anger. New York, NY: Basic Books; 1973.

[59] Bowlby, J. Attachment and Loss, Volume Three: Loss, Sadness and Depression. New York, NY: Basic Books; 1980a.

[60] Hartman, H. (1964) *Ego Psychology and the Problem of Adaptation*. New York: International Universities Press.

[61] Freud, S. (1959) The Ego and the Id. (A. Strachey, Trans. And J. Strachey, Ed.). New York: NY. (Original work published 1923).

[62] ten Have-de Labije, J., Neborsky, R.L.. 2012. Mastering Intensive Short-Term Dynamic Psychotherapy: A Roadmap to the Unconscious. London: Karnac Books.

[63] Purves, D., Augustine, G.J., Fitzpatrick, D., Hal, W.C., LaMantia, A.S., White, L.E. Neuroscience. 5th ed. 2012. 759 p.

[64] Price, R.H., et al. Principles of Psychology. 2nd ed. IL: Scott, Foresman and Co.; 1987. 624 p.

[65] Shah, S. G., Klumpp, H., Angstadt, M., Nathan, P. J., & Phan, K. L. 2009. Amygdala and insula response to emotional images in patients with generalized social anxiety disorder. *Journal of Psychiatry & Neuroscience: JPN, 34*(4), 296–302.

[66] Sharief AA[1], Badea A, Dale AM, Johnson GA. (2008). Automated segmentation of the actively stained mouse brain using multi-spectral MR microscopy. Neuroimage. 1;39(1):136–45.

[67] Tye, K.M., Prakash, R., Kim, S.-Y., Fenno, L.E., Grosenick, L., Zarabi, H., Thompson, K.R., Gradinaru, V., Ramakrishnan, C., Deisseroth, K. Amygdala circuitry mediating reversible and bidirectional control of anxiety. Nature. 2011;471(7338):358–362. DOI: 10.1038/nature09820.

[68] Tovote, P., Fadok, J.P., Lüthi, A. Neuronal circuits for fear and anxiety. Nature Reviews Neuroscience. 2015;16:317–331. DOI: 10.1038/nrn3945.

[69] Calhoon, G.G. & Tye, K.M. (2015). Resolving the neural circuits of anxiety. Nature Neuroscience, 18, 10.

[70] Jennings, J.H., Sparta, D.R., Stamatakis, A.M., Ung, R.L., Pleil, K.E., Kash T.L., Stuber, G.D. Distinct extended amygdala circuits for divergent motivational states. Science. 2013;496:224–228 . DOI: 10.1038/nature12041.

[71] Gray, J.A., McNaughton, N. The Neuropsychology of Anxiety: An Enquiry into the Functions of the Septo-Hippocampal System. 2nd ed. 2000.

[72] Malan, D. Individual Psychotherapy and the Science of Psychodynamics. London: Butterworth; 1979.

[73] Thayer, J.F., Lane, R.D. A model of neurovisceral integration in emotion regulation and dysregulation. Journal of Affective Disorders. 2000;61(3):201–216.

[74] Jänig, W., McLachlan, E.M. Neurobiology of the autonomic nervous system. In: Mathias, C.J., Bannister, R., editors. Autonomic Failure. 5th ed. Oxford: Oxford University Press; 2013. p. 21–34.

[75] Canteras, N.S., Swanson, L.W. Projections of the ventral subiculum to the amygdala, septum, and hypothalamus: a PHAL anterograde tract-tracing study in the rat. The Journal of Comparative Neurology. 1992;324(2):180–194. DOI: 10.1002/cne.903240204.

[76] Hayes S.C. Pistorello, J., Levin, M.E. 2012. Acceptance and Commitment Therapy as a Unified Model of Behavior Change. The Counseling Psychologist 40(7) 976–1002.

[77] Benjamin, L.S. Interpersonal Diagnosis and Treatment of Personality Disorder. 1st ed. New York, NY: Guilford Press; 1993.

[78] Davanloo, H. (2000). *Intensive short-term dynamic psychotherapy: Selected papers of Habib Davanloo*. Chichester, England: John Wiley & Sons.

[79] Beck, A.T., Emery, G., Greenberg, R.L. Anxiety Disorders and Phobias: A Cognitive Perspective. 1985.

[80] Grawe, K. 2007. Neuropsychotherapy. How the neurosciences inform effective psychotherapy. New York, NY: Psychology Press.

[81] Shedler, J. The efficacy of psychodynamic psychotherapy. American Psychologist. 2010;65(2):98–109. DOI: 10.1037/a0018378.

[82] Johansson, R, Town, JM, Abbass, A. (2014) Davanloo's Intensive Short-Term Dynamic Psychotherapy in a tertiary psychotherapy service: overall effectiveness and association between unlocking the unconscious and outcome. Peer J 2:e548. https://doi.org/10.7717/peerj.548

[83] Mennin, D.S., Fresco, D.M. Emotion regulation as an integrative framework for understanding and treating psychopathology. In: Kring, A.M., Sloan, D.M., editors. Emotion Regulation in Psychopathology: A Transdiagnostic Approach to Etiology and Treatment. New York, NY: Guildford Press; 2009. p. 356–379.

[84] Berking, M., Schwarz, J. The affect regulation training. In: Gross, J.J., editor. Handbook of Emotion Regulation. New York, NY: Guilford Press; 2014.

Chapter 2

Benzodiazepines and Anxiety Disorders: From Laboratory to Clinic

Janko Samardzic and Dubravka Svob Strac

Additional information is available at the end of the chapter

http://dx.doi.org/10.5772/64959

Abstract

Benzodiazepines (BDZs), which are among the most widely prescribed drugs in current psychiatric practice, act as positive modulators of GABAergic neurotransmission. They are used to treat a wide range of disorders, from anxiety, affective disorders and insomnia to epilepsy, alcohol withdrawal and muscle spasms. However, the development of tolerance and dependence after long-term BDZ treatment, as well as the abuse potential, limit their use. Although some other classes of drugs are currently considered as a better choice for long-term treatment, BDZs to date still remain indispensable drugs. They are widely prescribed for anxiety disorders, with high levels of evidence existing for the short-term BDZ use in panic disorder and generalized anxiety disorder, intermediate for social anxiety and poor in post-traumatic stress disorder and obsessive-compulsive disorder. Future studies are intending to develop the new selective drugs that act via BDZ receptors, but with novel, narrow profile of action. Furthermore, the research on alternative therapeutic approaches of psychiatric disorders has shifted the focus onto therapeutic potential of natural BDZ ligands.

Keywords: GABA, benzodiazepines, anxiety disorders, pharmacology, behaviour

1. Introduction

It was midst of twentieth century when the pharmaceutical industry recognized the growing need of modern society for various kinds of tranquilizers. The first real breakthrough was achieved with a group of drugs called barbiturates, which were initially considered to be almost omnipotent and safe [1]. However, with the growth of their use, the adverse effects of barbiturates have been painfully discovered—the development of drug addiction, a risk of overdose and increased number of suicides. Therefore, the intensive search for a better sedative has been continued. In the mid-fifties Dr. Leo Sternbach and his team from the company Hoffman-La Roche came into the focus of scientific publicity [2]. As it often happens in science, significant breakthroughs occur as a result of dedicated professional work and partly due to accidental discoveries. Working on a research of various molecules with possible sedative effects, Dr. Sternbach decided to re-examine benzheptoxidiazines, the group of substances that he worked on in Poland 20 years ago. Nearly 2 years later, while cleaning the warehouse, one of his assistants came across the substance left in storage for possible later testing, and Dr. Sternbach decided to examine it, just in case. It was the first 1,4-benzodiazepine, called chlordiazepoxide, demonstrating anxiolytic, hypnotic and muscle relaxant properties and only minor side effects, which has been marketed in a pharmaceutical form as Librium. Encouraged by the success of chlordiazepoxide, Hoffman-La Roche continued the research on related substances, which resulted in the synthesis of diazepam, an anxiolytic 3–10 times more potent than its precursor. Over 40 different benzodiazepine derivatives were synthesized in the following decades.

The next great success was the synthesis of triazolo-benzodiazepine analogues, such as alprazolam. Alprazolam is a specific anxiolytic in many ways—primarily because of its complex chemical structure, but also for its combined anxiolytic and antidepressant effects, which makes it useful for the specific cases of neuroses accompanied by depressive behaviour. However, in spite of extensive research, it took many years for researchers to associate benzodiazepines (BDZs) and their effects on γ-aminobutyric acid (GABA) and to propose their mechanism of action [3]. The mechanism of action of BDZ was established only 15 years after their introduction into clinical practice, during sixties [4], whereas 20 years passed until complete revelation of complex architecture of BDZ receptors [5]. Discovery of β-CCE, the first BDZ inverse agonist, in 1980, proved to be crucial for the future development of the benzodiazepine pharmacology, because it supported the idea of the existence of BDZ binding site subtypes in specific brain regions, which could have different physiological functions [6]. In contrast to other sedative-hypnotic drugs such as barbiturates, there is a specific antagonist for BDZ—flumazenil, which acts as a competitive antagonist in the presence of BDZ agonist compounds [7].

2. GABA and the brain

The inhibition and excitation of neural networks form the basis of information transfer in mammalian central nervous system (CNS). In an adequate balance between inhibitory and excitatory actions of neurotransmitters lies the key of normal functioning of the most complex processes in the brain [8]. The mutual coordination of main inhibitory neurotransmitter, GABA, and the major excitatory neurotransmitter, glutamate, is responsible for an adequate rhythmic activity, of both single as well as of group of neurons, thus altering synaptic plasticity and ensuring the normal functioning of CNS [9]. Decreased or increased activity of one or the other system is associated with number of neurological and psychiatric diseases [10, 11]. In addition, it has been suggested that the activity of inhibitory interneurons, most of which are GABAergic, defines the spatiotemporal framework, necessary for the different patterns of neural oscillations, which are essential for information processing in different brain structures

http://dx.doi.org/10.5772/64959

[12–14]. GABA was discovered in the 1950 by two independent research groups in the brain of mice, and it took additional 20 years to be officially proclaimed as a neurotransmitter [15–17]. The importance of GABA in neurotransmission is depicted by the fact that every third chemical synapse in the brain uses GABA as a neurotransmitter [18]. GABAergic synapse is the site of action of several different classes of drugs that modulate inhibitory neurotransmission [19]. Among these drugs are BDZs that have a wide spectrum of indications and represent a gold standard in treatment of anxiety disorders [20–22]. In the mid-seventies, the relationship between GABAergic system and anxiolytic action of BDZ was revealed, demonstrating that BDZ facilitates GABA neurotransmission. As positive modulators of GABAergic neurotransmission, BDZs have been described and examined in detail.

GABA achieves its effects by acting via two types of receptors: ionotropic GABA-A and metabotropic GABA-B receptors [23]. Although the third type for GABA receptor has been identified, and has its pharmacological specificities, the term GABA-C has not received broad consensus among experts, and IUPHAR (The International Union of Basic and Clinical Pharmacology) has classified it within GABA-A receptors [24]. GABA receptors, which are mainly coupled to the chloride channel but in varying degrees, can also couple to calcium, sodium and potassium channels. GABA-A receptors mediate the majority of GABA inhibitory actions in the CNS [10]. GABA-B receptors are localized pre- and post-synaptically, and negatively modulate adenylyl cyclase and inositol triphosphate synthesis, with final effect of potassium activation and/or inhibition of voltage-dependent calcium channels. Depending on the localization of GABA-B receptors, GABA-mediated inhibitory influences can be potentiated (post-synaptic receptors, presynaptic heteroreceptors on glutamergic endings) or reduced (autoreceptors) [24, 25].

GABA-A receptor is a pentameric complex made of transmembrane proteins that form ion channel, selectively permeable to chloride anion. These ligand-gated receptors are assembled from various (α1–6, β1–3, γ1–3, δ, ε, π, θ) subunits and their functional and pharmacological properties depend on their subunit composition [19, 26], GABA-A receptors are mainly localized in synapses, on post-synaptic membrane. In various regions, they are also localized extrasynaptically, especially when it comes to GABA-A receptor complex containing $\alpha 4$, $\alpha 5$ or $\alpha 6$ subunit [27, 28]. GABA-A receptors are also present on glial cells, and they could play an important role in adaptation of these cells to the needs of surrounding neurons [29]. GABAergic mechanisms are also involved in metabolic processes [19]. Negative correlation between the intensity of GABAergic neurotransmission and metabolic processes in cerebral tissue has been established [30, 31]. Activation of GABA-A receptors leads to a change in conformational state of associated ion channel, which results in an increased permeability to chloride ions [19, 24]. There are 14 different, structurally specific binding sites determined at the GABA-A receptor complex (**Figure 1**).

In addition to benzodiazepine binding site, i.e. BDZ receptor, at least 13 different, structurally specific sites on GABA-A receptor complex have been identified: (1) GABA and other agonists binding site, as well competitive antagonists; (2) picrotoxin site close to ion channel; (3) barbiturates binding site; (4) neuroactive steroids binding site; (5) ethanol binding site; (6) inhalation anaesthetics stereoselective binding sites; (7) furosemide diuretic binding site; (8)

Zn^{2+} ion binding site; (9) other divalent cations binding site; (10) La^{3+} ions site; (11) phosphorylation of specific protein kinases sites; (12) phospholipids binding sites; (13) sites involved in interaction of GABA-A receptor and microtubules, which take part in receptor grouping on post-synaptic membranes [32]. Via these binding sites, different modulators of GABA-A receptor complex exert their effects, acting as positive allosteric modulators, that potentiate the GABAergic effects, as negative modulators, that reduce the effects of GABA, and as neutral allosteric modulators, that competitively block the effects of these two types of agonists—antagonists [33]. Partial agonists and partial inverse agonists do not show full positive or negative modulation, even in the highest concentrations.

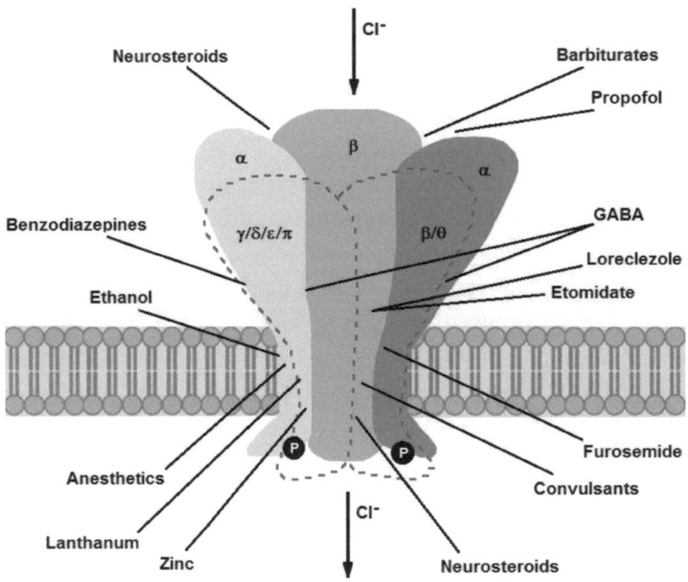

Figure 1. Schematic representation of GABA-A receptor complex and its subunits.

3. GABA-A receptor complex: a therapeutic target for BDZs

BDZ and other non-benzodiazepine analogues that bind to BDZ regulatory site (BDZ receptor), placed at interface between α and γ subunits, allosterically increase GABA receptor affinity. The result of this modulatory influence is increased opening frequency of ion channel in the presence of given neurotransmitter concentration, i.e. increased efficiency of GABAergic neurotransmission [34]. Although the $\gamma 2$ subunit presence is necessary, the selectivity by which BDZ ligands bind to GABA-A receptors of different structure primarily depends on six different α subunits. The functional and pharmacological properties of GABA-A receptors depend on their subunit composition [19]. Thus, $\alpha 1$ GABA-A receptors mediate sedative, amnesic and partly anticonvulsant, but not anxiolytic action of diazepam, in which $\alpha 2$ subunit

http://dx.doi.org/10.5772/64959

plays the major role [35]. That is how zolpidem has high binding affinity to GABA-A receptors containing $\alpha1$ subunit [35, 36]. Today, GABA-A receptors containing $\alpha1$ subunit are mainly marked as GABA-A1 receptors. 1,4-benzodiazepines, such as diazepam, with nearly same affinity bind to GABA-A1 receptors as to GABA-A receptors containing $\alpha2$, $\alpha3$ and $\alpha5$ subunits, marked as GABA-A2, GABA-A3 and GABA-A5 receptors. Therefore, diazepam shows comparable affinity to all BDZ-sensitive receptors (GABA-A receptors containing $\gamma2$, β, $\alpha1$, $\alpha2$, $\alpha3$ or $\alpha5$ subunit), while it does not bind to BDZ-insensitive receptors (receptors containing $\alpha4$ or $\alpha6$ subunit). On one GABA-A receptor complex, there is only one BDZ recognition site; although by rule this protein complex contains two α subunits [37]. Namely, γ subunit via specific amino acid residues takes part in BDZ receptor formation and is most probably placed in the presence of only one α subunit, so that the possibility of another BDZ molecule binding via another α subunit remains unaccomplished. On the other hand, GABA binds to receptor at the interface between α and β subunits, which are present with two pairs of neighbouring macromolecules, so that two GABA molecules can bind to one receptor complex.

Electrophysiological researches indicate that BDZ, applied at nanomolar concentrations, leads to conformational changes of one GABA binding site, most probably the one whose forming includes α subunit that was also involved in BDZ binding. Therefore, thanks to conformational change transfer from the α-γ interface (BDZ receptor) to the neighbouring α-β interface (GABA binding site), BDZ increases the binding affinity of one GABA molecule. Conformational changes do not transfer to other α subunit, i.e. to another GABA molecule binding site, which could explain the fact that BDZ receptors cannot by themselves open receptor ion channel, and that, unlike barbiturates, they cannot act in the absence of GABA [37].

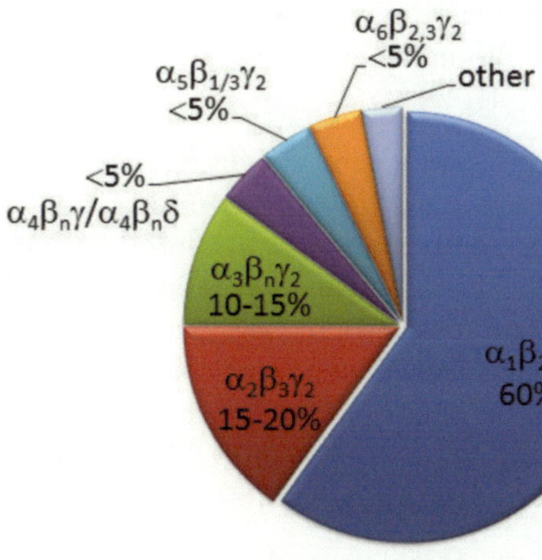

Figure 2. The distribution of different GABA-A receptor subtypes in the brain.

Receptors containing $\alpha1$, $\alpha2$, $\alpha3$ or $\alpha5$ subunit, in combination with $\gamma2$ or some of the β subunits, represent dominant fraction of GABA-A receptors and $\alpha1\beta2\gamma2S$ is the most common combination found in the brain (**Figure 2**) [24, 36]. Around 60% of all GABA-A receptors contain $\alpha1$ subunit. Regions in which these receptors are absent are rare; an example are motor neurons in spinal cord, but the distribution in certain regions is very unequal [36, 38]. In amygdaloid complex, for example, $\alpha1$ subunit is dominantly present in basolateral nucleus, and rarely found in the central nucleus [39]. The GABA-A receptors containing $\alpha2$ subunit represent 15–20% of entire population [36]. In small number of structures, this subunit is completely absent, and it is in general less expressed than $\alpha1$ subunit. One of the exceptions is central nucleus of amygdala, in which $\alpha2$ subunit is dominant [39]. Receptors containing this subunit are densely distributed at axon initial segments of cerebral cortex and hippocampus projection cells [35, 40]. The receptors with $\alpha3$ subunit are found in about 10–15% of the whole GABA-A receptor number [36]. The distribution is unequal, and in a number of regions these receptors are absent [38]. Less than 5% of GABA-A receptors contains $\alpha5$ subunit. In certain structures, these receptors are more expressed (certain regions of cerebral cortex, olfactory system, hypothalamus, basal ganglia) [38], while in hippocampus $\alpha5$ subunit is present in about 15–20% of all BDZ-sensitive GABA-A receptors [36, 41–43].

4. BDZs in clinical practice

BDZs are nowadays among the most widely prescribed drugs in current psychiatric practice and have relatively favourable pharmacokinetics and pharmacodynamic profile. Pharmacological profile of BDZ includes sedation, anxiolysis, hypnotic effect, anterograde amnesia, muscle relaxation and anti-convulsant properties. Different BDZs are available in practice, and basic indication of each BDZ preparation is declared by its most manifested effect. According to the guidelines and recommendations, systematic reviews and meta-analyses, there is a spectrum of use of BDZ in psychiatric clinical practice: anxiety and affective disorders, alcohol withdrawal, sleep disorders, delirium, aggressive behaviour in psychoses and neuroleptic-induced disorders (**Figure 3**) [2, 44, 45].

The main advantage of BDZs is their quick action, which can be seen soon after the first taking of the drug. BDZs are usually taken orally or may be administered intravenously. Due to well absorbance, they usually reach their maximum of concentration in plasma for about 1 h. They bind strongly to plasma proteins, and their high liposolubility causes many of them to accumulate gradually in the fatty tissue. BDZs are generally metabolized and excreted as glucuronides in urine [46]. According to their pharmacokinetics properties, BDZs are divided into three groups: short-acting (triazolam and midazolam), intermediate-acting (alprazolam, clonazepam, lorazepam, nitrazepam) and long-acting (diazepam, chlordiazepoxide, flurazepam).

They are mostly non-toxic and due to their large therapeutic range, BDZs are considered to be safer drugs than other medicaments of similar function (e.g. barbiturates). However, BDZ should be avoided or administered at lower dosages in the elderly, used sparingly in children, and applied with caution in the first and third trimesters of pregnancy and while breastfeeding [47]. Taken in excessive doses, they may lead to death very rarely, but we should bear in mind that the risk of toxic effects is increased in the presence of other CNS depressants, especially alcohol. The main side effects are drowsiness, confusion, anterograde amnesia and impaired motor function. Tolerance occurs following prolonged treatment with all benzodiazepines [48, 49], as well as an addiction which is their main drawback. Tolerance to the sedative effects develops already after relatively short period of exposure to BDZ, while the development of tolerance to the anxiolytic effects of these drugs require prolonged treatments. Tolerance to the BDZ anticonvulsant activity develops at medium speed. The speed of development of tolerance also depends on the BDZ dose administered. According to the efficacy of these drugs, tolerance and dependence develop to a lesser extent following treatment with partial BDZ agonist than after administration of full benzodiazepine agonists. Because of the possible development of addiction and withdrawal syndrome, it is advisable to limit BDZ therapy to a maximum of 4 weeks continuously, and then implement a gradual dose reduction [50]. The risk-benefit ratio remains positive in most patients in the short term, but is un-established beyond that time [51]. It should be also noted that occasionally they can cause a paradoxical effect, such as increased anxiety, excitement, irritability and aggressive behaviour.

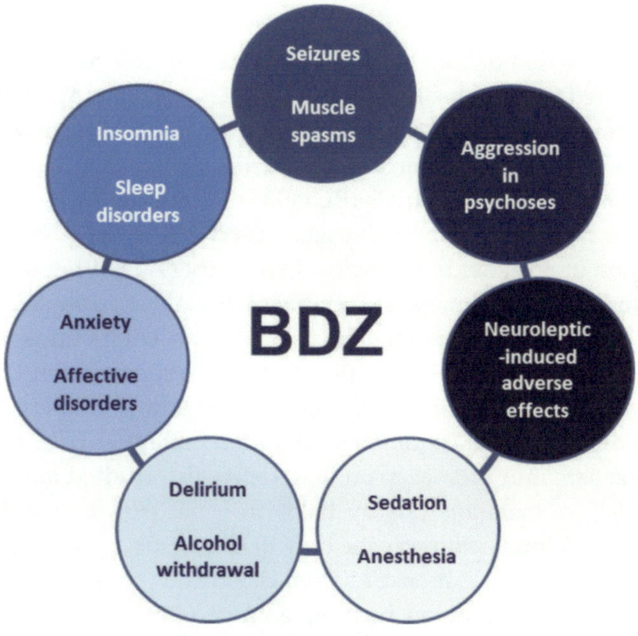

Figure 3. The use of BDZs in psychiatric clinical practice.

4.1. BDZs and anxiety disorders

According to the international guidelines, selective serotonin reuptake inhibitors (SSRIs) and serotonin/norepinephrine reuptake inhibitors (SNRIs) are currently recommended as the drugs of first choice for treatment of anxiety disorders, suggesting BDZ for the second-line or follow-up therapy [44, 45, 52]. Nevertheless, some recent studies have demonstrated that long-term use of BDZ can be effective and safe and that BDZ can be combined with antidepressants and psychological therapy to produce optimal outcomes [53]. Such findings have given an incentive to reconsider the role of BDZ for therapy of anxiety disorders. Due to rapid onset, BDZs are effective especially in the acute phase of anxiety. However, in addition to the side effects, one of the main disadvantages is that BDZs have no confirmed antidepressant activity, since comorbidity between anxiety and depression becomes the rule rather than the exception [44, 45, 52]. Although, alprazolam is internationally registered for the treatment of anxiety associated with depression, evidence of its specific antidepressant effect as a single treatment is inconclusive. In an extensive intervention review by van Marwijk et al. [54], alprazolam appears to reduce depressive symptoms more effectively than placebo and as effectively as tricyclic antidepressants. However, the studies included in this review were heterogeneous, of poor quality and only addressed short-term effects, thus limiting confidence in these findings. Here, we summarize the current available information related to the BDZ use for the most common anxiety disorders: generalized anxiety disorder, social anxiety disorder, panic and post-traumatic stress disorder.

4.1.1. Generalized anxiety disorder

Due to slower onset of action of SSRIs and SNRIs, BDZs are still recommended for the acute treatment of generalized anxiety disorders (GAD), either concomitantly until the effects of the antidepressant become apparent, or as a short-term measure for increased anxiety [47]. The efficacy and safety of BDZs in GAD, particularly for alprazolam and diazepam, were assessed in numerous randomized controlled trials (RCT) [55–67]. Bandelow et al. [44] recommended alprazolam in a dose between 1.5 and 6 mg when used adjunctively with an antidepressant and World Federation of Societies of Biological Psychiatry (WFSBP) guidelines also included the use of diazepam and lorazepam for the treatment of GAD in adult daily doses of 5–15 and 2–8 mg, respectively. WFSBP guidelines have rated both compounds with a category of evidence A for GAD treatment. However, the overall recommendation grade is lower, since the long-term treatment studies with BDZs in GAD are lacking and these compounds should only be used when other drugs or cognitive behavioural therapy (CBT) have failed [44]. Other BDZs, such as lorazepam and bromazepam, have been also studied and reviewed, but with less evidence [68, 69]. A meta-analysis by Baldwin et al. [70] reviewed 27 RCTs of drug treatments for GAD, and lorazepam was the only BDZ included. Fluoxetine has been ranked first drug of choice according to response and remission and pregabalin for tolerability, while lorazepam efficacy has been found low but with a limited data.

4.1.2. Social anxiety disorder

Effective treatment of social anxiety disorder (SAD) includes CBT and a spectrum of medications including some antidepressants, BDZs and anticonvulsants [71]. Many trials have been conducted with BDZs in the treatment of SAD. Munjack et al. [72] showed superior clinical efficacy of clonazepam for the patients with SAD, while Gelernter et al. [73] demonstrated a modest effect for alprazolam. Subsequently, Davidson et al. [74] demonstrated clinical benefits of clonazepam treatment for SAD, fear and phobic avoidance, inter-personal sensitivity, and on disability measures. Otto et al. [75] found that clonazepam or CBT were equally effective in acute treatment, while some authors found mixed results [2]. Anyhow, the reference data indicate BDZs to be efficacious treatments in SAD, well tolerated and with rapid effect; however, not as first-line treatment because of their potential withdrawal difficulties and limited spectrum of action. Thus, for SAD therapy, BDZs are rated with a category of evidence B by the WFSBP guidelines [44].

4.1.3. *Panic disorder*

Over the past 30 years, BDZs have been successfully used to treat the core symptoms of panic disorder (PD). Namely, the most robust evidence of BDZ efficacy in the treatment of anxiety was determined in panic disorder. BDZs are generally shown to be effective for a broad range of PD symptoms, with their rapid and maintained effect over a 7- to 8-month period [76]. The RCTs have demonstrated that alprazolam, diazepam, clonazepam and lorazepam are all clinically effective in PD. One of the first RCTs confirming the effect of BDZs in PD was performed with patients who were randomized to placebo group or alprazolam treatment for 8 weeks [77]. The efficacy of alprazolam was clearly demonstrated in patients usually consequently treated with tricyclic antidepressants or MAO inhibitors. In a following study, results indicated that alprazolam and diazepam appeared equally effective in patients with panic attacks and generalized anxiety compared to placebo [78]. Many trials subsequently attested the short-term efficacy of BDZ in relieving the core symptoms of PD. Thus, BDZs are rated for PD treatment with category of evidence A by the WFSBP guidelines, and this rate is consistent with their acute efficacy [44]. However, sufficient efficacy in PD is still lacking for the long-term BDZ therapy.

4.1.4. *Post-traumatic stress disorder*

The use of BDZ in post-traumatic stress disorder (PTSD) is poorly supported by current literature, and only a few available studies demonstrate mixed results. In the study by Braun et al. [79], 5 weeks of alprazolam treatment, showed only minimal improvement in the core symptoms of PTSD. In a following study, results indicated that alprazolam and clonazepam, after 6 months of treatment, appeared equally effective in patients with PTSD compared with control subject [80]. Moreover, some preclinical studies showed even increased vulnerability to stress after alprazolam single application in the animal model [81]. Thus, use of BDZ as a monotherapy in PTSD is not currently supported (category of evidence F by the WFSBP guidelines) [44]; however, potential benefits of combined treatment remain to be further elucidated. Similarly, mixed/negative results were found for efficacy of BDZ treatment in obsessive-compulsive disorder (OCD) [2, 82].

5. The perspective of BDZ research and development

It has been clearly shown that classic benzodiazepines, such as diazepam, achieve their effects by enhancing GABA neurotransmitter activity at the number of GABA-A receptors (BDZ-sensitive receptors), while they do not bind to the rest of receptor population [11]. At the beginning of the twenty-first century, using technology of genetic engineering, the conditions were created for establishing the correlation between certain effects of BDZ receptor ligands and their specific molecular and neural substrate of action [83]. Nowadays, these highly specific studies will probably result in the development of selective drugs that act via BDZ receptor with novel, narrow profile of action—therapeutic as well as adverse. Furthermore, the discovery and synthesis of a number of natural BDZ ligands have been initiated.

5.1. Pharmacokinetic and pharmacodynamic modifications

Pharmacokinetic modifications represent favourable approach since the substances with similar pharmacological activity are adapted to different therapeutic indications. The application of different BDZ as anxiolytics, hypnotics or myorelaxants, is primarily determined by their pharmacokinetic properties, such as dosage form, administration route and the presence of active metabolites. By reducing time of their action, great success has been achieved in the treatment of insomnia, thereby avoiding the morning drowsiness, and enabling the motor vehicle operation [84]. However, this type of modifications could not solve problems such as cognitive impairments and loss of coordination, especially in elderly. Since pharmacokinetic modifications did not lead to separation of desired and unwanted effects of BDZ, new approaches applied to the synthesis of a number of substances with different pharmacological profiles.

Along with pharmacokinetic modifications, a new concept emerged with the assumption that BDZ with lower efficiency at GABA-A receptors could maintain the useful anxiolytic effect of standard BDZ, but without unwanted side effects [85]. By using this concept, bretazenil was developed, which has not shown *in vitro* selectivity for certain subtypes of GABA-A receptor [86, 87]. Although acting as partial agonist of BDZ receptors, bretazenil has been shown to be potent anxiolytic and anticonvulsant in animal models [86, 88, 89]. Despite of bretazenil anxiolytic effect, noticed in the human studies, clinical studies were discontinued in the first phase due to deep sedation, whereby the sedative effect of bretazenil even at lower doses could be compared to the effects of diazepam and ethanol combination. Sedative effect of bretazenil was so pronounced that certain responders fell asleep while performing routine tests [90]. Further research in the field of non-selective partial agonists of BDZ receptors have led to the synthesis of imidazenil, a substance with 30–50% lower efficiency than diazepam [91]. Although imidazenil is more potent than bretazenil and practically without any potential for inducing sedation in preclinical models, studies on humans have not confirmed the expected results [92, 93]. It is important to mention that imidazenil showed a lower efficiency at $\alpha 1$ GABA-A receptors in relation to diazepam, which could be one of the reasons of such favourable pharmacological profile, but also indicated new possibilities in the search for ideal anxiolytic drug [94]. However, after several unsuccessful attempts, the idea of partial agonist development was not so attractive any more, especially in the light of discovery that there are several possible subtypes of GABA-A receptor complex. As a result, the interest in the new drug development has been directed towards selective ligands [36].

5.2. The development of BDZ ligands selective for certain receptor subtypes

Diazepam, synthesized in 1963, for many years represented a synonym for BDZ group of drugs, and only later a huge number of BDZ ligands were synthesized. Diazepam has a wide spectrum of indications and shows anxiolytic, hypnotic, myorelaxant, anticonvulsant and other effects. Moreover, the examination of diazepam, as a reference BDZ, on genetically modified animals provided linkage of certain effects with specific receptor subtype and enabled further development of selective drugs with new, narrow profiles of action [35, 95, 96]. The past decade of research has led to an increased understanding of the specific GABA-A receptor subtypes responsible for various pharmacological effects of BDZs (**Table 1**) [97, 98].

Effect	Receptor subtypes			
	α_1	α_2	α_3	α_5
Sedation	+	–	–	–
Anxiolytic activity	–	+	–	–
Anterograde amnesia	+	–	–	+
Myorelaxation	–	+	+	+
Anticonvulsive activity	+	–	–	–
Addiction	+	–	–	–

Table 1. Pharmacological effects of BDZs and the corresponding subtypes of GABA-A receptors.

So far, 19 subunits of GABA-A receptor complex have been cloned, and subunits are classified into several structurally connected subfamilies comprising highly homologous isoforms (α1–6, β1–3, γ1–3, δ, ε, θ, π, ϱ1–3). The most commonly represented receptor population is a complex consisted of two α and two β and one γ subunit [10, 99]. Concerning the fact that two different α and/or β subunit can be present in the same receptor, in the brain, it is highly probable that there are more than 500 different GABA-A receptor subtypes. However, molecular studies show that the number of main subtypes of GABA-A receptors is less than 10, with the $\alpha1\beta2\gamma2$ present in about 43%, $\alpha2\beta2\gamma3$ in 18%, and $\alpha3\beta\gamma2/3$ in 17% of receptors [100, 101]. On the other hand, extensive immunochemical studies which investigated the distribution of 13 most common subunits (α1–6, β1–3, γ1–3, δ) in the rat brain, have shown that the great majority of receptors contain γ2 subunit, while δ and γ1 subunits are confined to a very small number of brain regions. Among α subunits, α1 showed the most pronounced immunoreactivity in practically all areas of the brain. Other α subunits are more confined to particular parts of the brain [38].

The development of ligands selective for α1 subunit (e.g. zolpidem and zaleplon) has been accomplished. Zolpidem represents selective agonist for GABA-A receptors containing α1 subunit, and it is widely used in the treatment of insomnia [102]. Zolpidem is characterized by 5–14 times greater affinity for α1 in relation to α2 and α3 receptor subtypes, as well as by very low affinity for α5GABA-A receptors [103]. Zaleplon is another α1GABA-A selective agonist from the so-called Z-hypnotics group, which in relation to zolpidem has two times lower affinity for α1GABA-A receptors, and unlike zolpidem shows certain affinity for α5GABA-A receptors [104]. Only sporadic reports can be found on behavioural effects of ligands characterized by selective affinity/efficiency at α2- and α3-subtype of GABA-A receptor complex, which in preclinical conditions have demonstrated anxiolytic action. For current pharmacological trials, particularly α5 subunit of GABA-A receptor represents the most interesting drug target, primarily from the aspect of new pro-cognitive drugs development. The results of the study with genetically modified mice have shown that decreased expression of α5 subunit acts facilitatory on performing certain memory tests, so over the next couple of years, various inverse agonists with selective affinity or efficiency at α5GABA-A receptors were synthesized [43, 105, 106]. L-665, 708, partial inverse agonist at all four subtypes of GABA-A receptors, with greater functional selectivity at α5GABA-A receptors, improves learning and motivation in rats during forced swimming [107]. In addition, α5IA is a substance with similar profile, which has also shown promnestic activity, as well as selective influence on the acquisition and retrieval of memory [108]. Finally, RO4938581, a selective inverse agonist at α5GABA-A receptors has successfully antagonized impairing effects of scopolamine and diazepam on rats learning in Morris water maze [109]. Clinical studies have shown that in healthy volunteers, pre-treatment with α5 selective inverse agonist has greatly decreased amnesic effect caused by alcohol intake, confirming the concept validity of α5GABA-A receptors significance in hippocampal-dependent memory processes [110]. However, more clinical studies remain to be conducted.

5.3. Natural compounds as BDZ receptor ligands

Extracts of many plants, as well as their natural and synthetic derivatives, are increasingly proposed as an integral part of the clinical treatment of psychiatric diseases, particularly mood disorders, due to higher compliance and fewer side effects in patients [111, 112]. One of the first studies on natural compounds as CNS ligands was reported by Roche researchers, in which they isolated a few *'diazepam-like'* isoflavan derivatives with low affinity for BDZ receptor [113]. Chrysin, the first flavonoid described as a specific partial agonist of BDZ receptor, is almost equipotent to diazepam as an anxiolytic, but does not exert myorelaxant and sedative effects [114]. In subsequent preclinical studies, the flavonoid cirsiliol showed a very low affinity for BDZ receptors and was devoid of anxiolytic actions, while kaempferol, quercetin and myricetin were active after oral application, exerting anxiolytic effects [115]. Apigenin, a common flavonoid found in many plants including chamomile, was initially described as a BDZ antagonist with anxiolytic effects [116]. However, later studies have showed that apigenin fits more into pharmacological profile of an inverse BDZ agonist, exerting sedative, but not anxiolytic effect [117]. Since this initial search for natural/alternative BDZ ligands, a significant number of successful derivatives have been isolated [112]. *In vitro* and *in vivo* studies have sufficiently clarified the pharmacological actions of flavonoid and flavonol derivatives on BDZ receptors; however, their precise mechanism seems to be much more complex.

Besides flavonoids, terpenoids have been also reported to affect GABA-A receptors in various ways, including modulation of BDZ binding site [118]. The most widely studied terpenoid is picrotoxin, a convulsant and non-competitive antagonist at GABA-A receptors. A structurally similar to picrotoxin is bilobalide, a terpenoid from the *Ginkgo biloba* plant, which acts as a negative allosteric modulator of GABA-A receptors, partially explaining cognition-enhancing effects of *Ginkgo biloba* extracts [119]. Studies have shown that bilobalide particularly modulate the peripheral BDZ receptors [120]. Furthermore, the extracts of *Valeriana* contain a large number of substances including terpenoids and flavonoids, many of which are considered to be active at GABA-A receptor complex [121]. More studies are still required in order to fully understand the pharmacological profile of natural CNS ligands and their potential in the treatment of neuropsychiatric disorders.

6. Conclusion

Thanks to a very wide range of pharmacological actions, including anxiolytic, sedative-hypnotic, anticonvulsant and myorelaxant effects, BDZs became the most frequently used group of psychoactive drugs in clinical medicine in the last 50 years. They have been used to treat a wide range of disorders, from anxiety, affective disorders and insomnia to epilepsy, alcohol withdrawal and muscle spasms. However, in spite of rapid onset of efficacy, relative safety and mild side-effects of these drugs, the knowledge about the development of tolerance and dependence after long-term BDZ treatment, as well as their abuse potential, has somewhat limited their previous widespread use. Although some other classes of drugs are currently considered as a better choice for long-term treatment, BDZs to date still remain indispensable drugs in clinical medicine. They are widely prescribed as a first line treatment in anxiety disorders, with high levels of evidence existing for the short-term BDZ use in PD and GAD treatment, intermediate for SAD therapy and poor in PTSD and OCD. Further research for new alternative drugs has been encouraged by the increased understanding of the mechanisms of BDZ action. Information and knowledge about the structure and function of GABA-A receptor complex provided the molecular basis for further elucidation of the nature of interactions between BDZs and their receptors. As a result, the selective substances, such as zopiclone, zolpidem and zaleplon, which also interact with the BDZ receptors, have been developed. Future studies will probably result in the development of new selective drugs that act via BDZ receptors with novel, narrow profile of action—therapeutic as well as adverse. Furthermore, the research on alternative therapeutic approaches of neuropsychiatric disorders has shifted the focus onto therapeutic potential of natural BDZ ligands.

Acknowledgements

Publication of this chapter was supported in part by Dunav Insurance Company A.D.O. Belgrade, Serbia. The authors declare that there are no competing interests regarding the publication of this chapter.

Author details

Janko Samardzic[1*] and Dubravka Svob Strac[2]

*Address all correspondence to: janko@med.bg.ac.rs

1 Institute of Pharmacology, Clinical Pharmacology and Toxicology, Medical Faculty, University of Belgrade, Belgrade, Serbia

2 Laboratory for Molecular Neuropsychiatry, Division of Molecular Medicine, Rudjer Boskovic Institute, Zagreb, Croatia

References

[1] Norn S, Permin H, Kruse E, Kruse PR. On the history of barbiturates. Dan Medicinhist Arbog. 2015;43:133–151.

[2] Dell'osso B, Lader M. Do benzodiazepines still deserve a major role in the treatment of psychiatric disorders? A critical reappraisal. Eur Psychiatry. 2013;28(1):7–20. DOI: 10.1016/j.eurpsy.2011.11.003.

[3] Wick JY. The history of benzodiazepines. Consult Pharm. 2013;28(9):538–548. DOI: 10.4140/TCP.n.2013.538.

[4] Costa E, Guidotti A, Mao CC. Evidence for the involvement of GABA in the action of benzodiazepins. Adv Biochem Psychopharmacol. 1975;14:113–130.

[5] Sieghart W. Structure and pharmacology of γ-aminobutyric acid, receptor subtypes. Pharmacol Rev 1995;47(2):181–234.

[6] Nielsen M, Braestrup C. Ethyl beta-carboline-3-carboxylate shows differential benzodiazepine receptor interaction. Nature. 1980;286(5773):606–607.

[7] Hood SD, Norman A, Hince DA, Melichar JK, Hulse GK. Benzodiazepine dependence and its treatment with low dose flumazenil. Br J Clin Pharmacol. 2014;77(2):285–294. DOI: 10.1111/bcp.12023.

[8] Petroff OA. GABA and glutamate in the human brain. Neuroscientist. 2002;8(6):562–573.

[9] Foster AC, Kemp JA. Glutamate- and GABA-based CNS therapeutics. Curr Opin Pharmacol. 2006;6(1):7–17.

[10] Jacob TC, Moss SJ, Jurd R. GABA(A) receptor trafficking and its role in the dynamic modulation of neuronal inhibition. Nat Rev Neurosci. 2008;9(5):331–343. DOI: 10.1038/nrn2370.

[11] Millan MJ. The neurobiology and control of anxious states. Prog Neurobiol. 2003;70(2): 83–244.

[12] Buzsaki G, Draguhn A. Neuronal oscillations in cortical networks. Science. 2004;304(5679):1926–1929.

[13] Klausberger T, Magill PJ, Marton LF, Roberts JDB, Cobden PM, Buzsaki G, Somogyi P. Brain state- and cell type-specific firing of hippocampal interneurons in vivo. Nature. 2003;421(6925):844–848.

[14] Paulsen O, Moser EI. A model of hippocampal memory encoding and retrieval: GABAergic control of synaptic plasticity. Trends Neurosci. 1998;21(7):273–278.

[15] Awapara J, Landua AJ, Fuerst R, Seale B. Free gamma-aminobutyric acid in brain. J Biol Chem. 1950;187(1):35–39.

[16] Roberts E, Frankel S. Gamma-aminobutyric acid in brain: its formation from glutamic acid. J Biol Chem. 1950;187(1):55–63.

[17] Roberts E. GABA: the road to neurotransmitter status. In: Olsen RW, Venter JC, editors. Benzodiazepines/GABA Receptors and Chloride Channels: Structural and Functional Properties. New York: Alan R. Liss, Inc; 1986. p. 1–39.

[18] Burt DR. Reducing GABA receptors. Life Sci. 2003;73(14):1741–1758.

[19] Korpi ER, Grunder G, Luddens H. Drug interactions at GABA(A) receptors. Prog Neurobiol. 2002;67(2):113–159.

[20] Smith TA. Type A gamma-aminobutyric acid (GABAA) receptor subunits and benzodiazepine binding: significance to clinical syndromes and their treatment. Br J Biomed Sci. 2001;58(2):111–121.

[21] Sramek JJ, Zarotksy V, Cutler NR. Generalized anxiety disorder: treatment options. Drugs. 2002;62(11):1635–1648.

[22] Stahl S. Don't ask, don't tell, but benzodiazepines are still the leading treatments for anxiety disorder. J Clin Psychiatry. 2002;63(9):756–757.

[23] Chebib M, Johnston GA. The 'ABC' of GABA receptors: a brief review. Clin Exp Pharmacol Physiol. 1999;26(11):937–940.

[24] Olsen RW, Sieghart W. International Union of Pharmacology. LXX. Subtypes of gamma-aminobutyric acid(A) receptors: classification on the basis of subunit composition, pharmacology, and function. Update. Pharmacol Rev. 2008;60(3):243–260. DOI: 10.1124/pr.108.00505.

[25] Bettler B, Kaupmann K, Mosbacher J, Gassmann M. Molecular structure and physiological functions of GABA(B) receptors. Physiol Rev. 2004;84(3):835–867.

[26] Barnard EA, Skolnick P, Olsen RW, Möhler H, Sieghart W, Biggio G, et al. International Union of Pharmacology. XV. Subtypes of gamma-aminobutyric acidA receptors:

classification on the basis of subunit structure and receptor function. Pharmacol Rev. 1998;50(2):291–313.

[27] Brunig I, Scotti E, Sidler C, Fritschy JM. Intact sorting, targeting, and clustering of gamma-aminobutyric acid A receptor subtypes in hippocampal neurons in vitro. J Comp Neurol. 2002;443(1):43–55.

[28] Fritschy JM, Brunig I. Formation and plasticity of GABAergic synapses: physiological mechanisms and pathophysiological implications. Pharmacol Ther. 2003;98(3):299–323.

[29] Lin SC, Bergles DE. Synaptic signaling between neurons and glia. Glia. 2004;47(3):290–298.

[30] Obradovic D, Bokonjic D, Savic M, Andelkovic D, Ugresic N, Stojiljkovic M. GABA-benzodiazepine receptor complex in brain oxidative metabolism regulation. Pharmacol Res. 2002;46(2):149–154.

[31] Obradovic D, Savic M, Andelkovic D, Ugresic N, Bokonjic D. The influence of midazolam and flumazenil on rat brain slices oxygen consumption. Pharmacol Res. 2003;47(2):127–131.

[32] Chebib M, Johnston GA. GABA-Activated ligand gated ion channels: medicinal chemistry and molecular biology. J Med Chem. 2000;43(8):1427–1447.

[33] Samardzic J, Svob Strac D, Obradovic M, Opric D, Obradovic DI. DMCM, a benzodiazepine site inverse agonist, improves active avoidance and motivation in the rat. Behav Brain Res. 2012;235(2):195–199. DOI: 10.1016/j.bbr.2012.07.032.

[34] Mehta AK, Ticku MK. An update on GABAA receptors. Brain Res Rev. 1999;29(2–3):196–217.

[35] Rudolph U, Crestani F, Möhler H. GABA(A) receptor subtypes: dissecting their pharmacological functions. Trends Pharmacol Sci. 2001;22(4):188–194.

[36] Möhler H, Fritschy JM, Rudolph U. A new benzodiazepine pharmacology. J Pharmacol Exp Ther. 2002;300(1):2–8.

[37] Williams DB, Akabas MH. Evidence for distinct conformations of the two alpha 1 subunits in diazepam-bound GABA(A) receptors. Neuropharmacology. 2001;41(5): 539–545.

[38] Pirker S, Schwarzer C, Wieselthaler A, Sieghart W, Sperk G. GABA(A) receptors: immunocytochemical distribution of 13 subunits in the adult rat brain. Neuroscience. 2000;101(4):815–850.

[39] Kaufmann WA, Humpel C, Alheid GF, Marksteiner J. Compartmentation of alpha 1 and alpha 2 GABA(A) receptor subunits within rat extended amygdala: implications for benzodiazepine action. Brain Res. 2003;964(1):91–99.

[40] Rudolph U, Möhler H. Analysis of GABAA receptor function and dissection of the pharmacology of benzodiazepines and general anesthetics through mouse genetics. Annu Rev Pharmacol Toxicol. 2004;44:475–498.

[41] Atack JR. Preclinical and clinical pharmacology of the GABA-A receptor alpha5 subtype-selective inverse agonist alpha5IA. Pharmacol Ther. 2010;125(1):11–26. DOI: 10.1016/j.pharmthera.2009.09.001.

[42] Harris D, Clayton T, Cook J, Sahbaie P, Halliwell RF, Furtmüller R, et al. Selective influence on contextual memory: physiochemical properties associated with selectivity of benzodiazepine ligands at GABAA receptors containing the alpha5 subunit. J Med Chem. 2008;51(13):3788–3803. DOI: 10.1021/jm701433b.

[43] Savic MM, Clayton T, Furtmüller R, Gavrilovic I, Samardzic J, Savic S, et al. PWZ-029, a compound with moderate inverse agonist functional selectivity at GABAA receptors containing $\alpha5$ subunits, improves passive, but not active avoidance learning in rats. Brain Res. 2008;1208:150–159. DOI: 10.1016/j.brainres.2008.02.020.

[44] Bandelow B, Zohar J, Hollander E, Kasper S, Moller HJ, on behalf of the WFSBP Task Force on Treatment Guidelines for Anxiety, Obsessive-Compulsive and Posttraumatic Stress Disorders. World Federation of Societies of Biological Psychiatry (WFSBP) Guidelines for the pharmacological treatment of anxiety, obsessive-compulsive and posttraumatic stress disorders – first revision. World J Biol Psychiatry. 2008;9(4):248–312. DOI: 10.1080/15622970802465807.

[45] Kennedy SH, Lam RW, Parikh SV, Patten SB, Ravindran AV. Canadian Network for Mood and Anxiety Treatments (CANMAT) Clinical guidelines for the management of major depressive disorder in adults. J Affect Disord. 2009;117(1):S5–S14. DOI: 10.1016/ j.jad.2009.06.044.

[46] Bennett PN, Brown MJ, Sharma P. Clinical Pharmacology. 11th ed. Edinburgh: Churchill Livingstone; 2012.

[47] Canadian Agency for Drugs and Technologies in Health. Short- and Long-Term Use of Benzodiazepines in Patients with Generalized Anxiety Disorder: A Review of Guidelines [Internet]. Ottawa (ON); 2014.

[48] Svob Strac D, Vlainic J, Jazvinscak Jembrek M, Pericic D. Differential effects of diazepam treatment and withdrawal on recombinant GABAA receptor expression and functional coupling. Brain Res. 2008;1246:29–40. DOI: 10.1016/j.brainres.2008.09.093.

[49] Vlainic J, Jembrek MJ, Vlainic T, Strac DS, Pericic D. Differential effects of short- and long-term zolpidem treatment on recombinant $\alpha1\beta2\gamma2s$ subtype of GABA(A) receptors in vitro. Acta Pharmacol Sin. 2012;33(12):1469–1476. DOI: 10.1038/aps.2012.89.

[50] Samardzic J. Case reports: Anxiety disorder. In: Prostran M, editor. Clinical Pharmacology. Belgrade: Cibid; 2012. p. 258–259.

[51] Lader M. Benzodiazepines revisited – will we ever learn? Addiction. 2011;106(12):2086–2109. DOI: 10.1111/j.1360-0443.2011.03563.x.

[52] Davidson JR, Zhang W, Connor KM, Ji J, Jobson K, Lecrubier Y, et al. A psychopharmacological treatment algorithm for generalised anxiety disorder (GAD). J Psychopharmacol. 2010;24(1):3–26. DOI: 10.1177/0269881108096505.

[53] Starcevic V. The reappraisal of benzodiazepines in the treatment of anxiety and related disorders. Expert Rev Neurother. 2014;14(11):1275–1286. DOI: 10.1586/14737175.2014.963057.

[54] van Marwijk H, Allick G, Wegman F, Bax A, Riphagen II. Alprazolam for depression. Cochrane Database Syst Rev. 2012;(7):CD007139. DOI: 10.1002/14651858.CD007139.pub2.

[55] Enkelmann R. Alprazolam versus buspirone in the treatment of outpatients with generalized anxiety disorder. Psychopharmacology (Berl). 1991;105(3):428–432.

[56] Hoehn-Saric R, McLeod DR, Zimmerli WD. Differential effects of alprazolam and imipramine in generalized anxiety disorder: somatic versus psychic symptoms. J Clin Psychiatry. 1988;49(8):293–301.

[57] Lydiard RB, Ballenger JC, Rickels K. A double-blind evaluation of the safety and efficacy of abecarnil, alprazolam, and placebo in outpatients with generalized anxiety disorder. Abecarnil Work Group. J Clin Psychiatry. 1997;58(11):11–18.

[58] Moller HJ, Volz HP, Reimann IW, Stoll KD. Opipramol for the treatment of generalized anxiety disorder: a placebo-controlled trial including an alprazolam-treated group. J Clin Psychopharmacol. 2001;21(1):59–65.

[59] Ansseau M, Olie' JP, von Frenckell R, Jourdain G, Stehle B, Guillet P. Controlled comparison of the efficacy and safety of four doses of suriclone, diazepam, and placebo in generalized anxiety disorder. Psychopharmacology (Berl). 1991;104(4):439–443.

[60] Boyer WF, Feighner JP. A placebo-controlled double-blind multicenter trial of two doses of ipsapirone versus diazepam in generalized anxiety disorder. Int Clin Psychopharmacol. 1993;8(3):173–176.

[61] Fontaine R, Annable L, Chouinard G, Ogilvie RI. Bromazepam and diazepam in generalized anxiety: a placebo-controlled study with measurement of drug plasma concentrations. J Clin Psychopharmacol. 1983;3(2):80–87.

[62] Rickels K, Case WG, Downing RW, Winokur A. Long-term diazepam therapy and clinical outcome. JAMA 1983;250(6):767–771.

[63] Rickels K, DeMartinis N, Aufdembrinke B. A double-blind, placebo-controlled trial of abecarnil and diazepam in the treatment of patients with generalized anxiety disorder. J Clin Psychopharmacol. 2000;20(1):12–18.

[64] Rickels K, Schweizer E, DeMartinis N, Mandos L, Mercer C. Gepirone and diazepam in generalized anxiety disorder: a placebo-controlled trial. J Clin Psychopharmacol. 1997;17(4):272–277.

[65] Feighner JP, Merideth CH, Hendrickson GA. A double-blind comparison of buspirone and diazepam in outpatients with generalized anxiety disorder. J Clin Psychiatry. 1982;43(12/2):103–108.

[66] Jacobson AF, Dominguez RA, Goldstein BJ, Steinbook RM. Comparison of buspirone and diazepam in generalized anxiety disorder. Pharmacotherapy 1985;5(5):290–296.

[67] Rickels K, Pollack MH, Feltner DE, Lydiard RB, Zimbroff DL, Bielski RJ, et al. Pregabalin for treatment of generalized anxiety disorder: a 4-week, multicenter, double-blind, placebo-controlled trial of pregabalin and alprazolam. Arch Gen Psychiatry. 2005;62(9):1022–1030.

[68] Feltner DE, Crockatt JG, Dubovsky SJ, Cohn CK, Shrivastava RK, Targum SD, et al. A randomized, double-blind, placebo-controlled, fixed-dose, multi-center study of pregabalin in patients with generalized anxiety disorder. J Clin Psychopharmacol. 2003;23(3):240–249.

[69] Llorca PM, Spadone C, Sol O, Danniau A, Bougerol T, Corruble E, et al. Efficacy and safety of hydroxyzine in the treatment of generalized anxiety disorder: a 3-month double-blind study. J Clin Psychiatry. 2002;63(11): 1020–1027.

[70] Baldwin D, Woods R, Lawson R, Taylor D. Efficacy of drug treatments for generalised anxiety disorder: systematic review and meta-analysis. BMJ. 2011;11:1199. DOI: 10.1136/bmj.d1199.

[71] Masdrakis VG, Turic D, Baldwin DS. Pharmacological treatment of social anxiety disorder. Mod Trends Pharmacopsychiatri. 2013;29:144–153. DOI: 10.1159/000351960.

[72] Munjack DJ, Baltazar PL, Bohn PB, Cabe DD, Appleton AA. Clonazepam in the treatment of social phobia: a pilot study. J Clin Psychiatry. 1990;51:35–40.

[73] Gelernter CS, Uhde TW, Cimbolic P, Arnkoff DB, Vittone BJ, Tancer ME, et al. Cognitive-behavioral and pharmacological treatments of social phobia. A controlled study. Arch Gen Psychiatry. 1991;48(10):938–945.

[74] Davidson JR, Potts N, Richichi E, Krishnan R, Ford SM, Smith R, et al. Treatment of social phobia with clonazepam and placebo. J Clin Psychopharmacol. 1993;13(6):423–428.

[75] Otto MW, Pollack MH, Gould RA, Worthington JJ, McArdle ET, Rosenbaum JF. A comparison of the efficacy of clonazepam and cognitive-behavioral group therapy for the treatment of social phobia. J Anxiety Disord. 2000;14(4):345–358.

[76] Davidson JR. Use of benzodiazepines in panic disorder. J Clin Psychiatry. 1997;58(2): 26–28.

[77] Chouinard G, Annable L, Fontaine R, Solyom L. Alprazolam in the treatment of generalized anxiety and panic disorders: a double-blind placebo-controlled study. Psychopharmacology (Berl). 1982;77(3):229–233.

[78] Dunner DL, Ishiki D, Avery DH, Wilson LG, Hyde TS. Effect of alprazolam and diazepam on anxiety and panic attacks in panic disorder: a controlled study. J Clin Psychiatry. 1986;47(9):458–460.

[79] Braun P, Greenberg D, Dasberg H, Lerer B. Core symptoms of posttraumatic stress disorder unimproved by alprazolam treatment. J Clin Psychiatry. 1990;51(6):236–238.

[80] Gelpin E, Bonne O, Peri T, Brandes D, Shalev AY. Treatment of recent trauma survivors with benzodiazepines: a prospective study. J Clin Psychiatry. 1996;57(9):390–394.

[81] Matar MA, Zohar J, Kaplan Z, Cohen H. Alprazolam treatment immediately after stress exposure interferes with the normal HPA-stress response and increases vulnerability to subsequent stress in an animal model of PTSD. Eur Neuropsychopharmacol. 2009;19(4):283–295. DOI: 10.1016/j.euroneuro.2008.12.004.

[82] Crockett BA, Churchill E, Davidson JR. A double-blind combination study of clonazepam with sertraline in obsessive-compulsive disorder. Ann Clin Psychiatry. 2004;16(3): 127–132.

[83] Rudolph U, Mohler H. Genetically modified animals in pharmacological research: future trends. Eur J Pharmacol. 1999;375(1–3):327–337.

[84] Mendelson WB. Clinical distinctions between long-acting and short-acting benzodiazepines. J Clin Psychiatry. 1992;53:4–7.

[85] Whiting PJ. GABA–A receptors: a viable target for novel anxiolytics? Curr Opin Pharmacol. 2006;6(1):24–29.

[86] Griebel G. Is there a future for neuropeptide receptor ligands in the treatment of anxiety disorders? Pharmacol Ther. 1999;82(1):1–61.

[87] Martin JR, Pieri L, Bonetti EP, Schaffner R, Burkard WP, Cumin R, et al. Ro16–6028: a novel anxiolytic acting as a partial agonist at the benzodiazepine receptor. Pharmacopsychiatry. 1988;21(6):360–362.

[88] Facklam M, Schoch P, Bonetti EP, Jenck F, Martin JR, Moreau JL, et al. Relationship between benzodiazepine receptor occupancy and functional effects in vivo of four ligands of differing intrinsic efficacies. J Pharmacol Exp Ther. 1992;261(3):1113–1121.

[89] Martin JR, Schoch P, Jenck F, Moreau JL, Haefely WE. Pharmacological characterization of benzodiazepine receptor ligands with intrinsic efficacies ranging from high to zero. Psychopharmacology. 1993;111(4):415–422.

[90] van Steveninck AL, Gieschke R, Schoemaker RC, Roncari G, Tuk B, Pieters MS, et al. Pharmacokinetic and pharmacodynamic interactions of bretazenil and diazepam with alcohol. Br J Clin Pharmacol. 1996;41(6):565–573.

[91] Giusti P, Ducic I, Puia G, Arban R, Walser A, Guidotti A, Costa E. Imidazenil: a new partial positive allosteric modulator of gamma-aminobutyric acid (GABA) action at GABAA receptors. J Pharmacol Exp Ther. 1993;266(2):1018–1028.

[92] Kadriu B, Guidotti A, Costa E, Davis JM, Auta J. Acute imidazenil treatment after the onset of DFP-induced seizure is more effective and longer lasting than midazolam at preventing seizure activity and brain neuropathology. Toxicol Sci. 2011;120:136–145. DOI: 10.1093/toxsci/kfq356.

[93] Wang Y, Oguntayo S, Wei Y, Wood E, Brown A, Jensen N, Auta J, Guiodotti A, Doctor BP, Nambiar MP. Neuroprotective effects of imidazenil against chemical warfare nerve agent soman toxicity in guinea pigs. Neurotoxicology. 2012;33(2):169–177. DOI: 10.1016/j.neuro.2011.12.018.

[94] Guidotti A, Auta J, Davis JM, Dong E, Grayson DR, Veldic M, et al. GABAergic dysfunction in schizophrenia: a new treatment target on the horizon. Psychopharmacology. 2005;180(2):191–205.

[95] Möhler H, Crestani F, Rudolph U. GABA(A)-receptor subtypes: a new pharmacology. Curr Opin Pharmacol. 2001;1(1):22–25.

[96] Collinson N, Atack JR, Laughton P, Dawson GR, Stephens DN. An inverse agonist selective for α5 subunit-containing GABAA receptors improves encoding and recall but not consolidation in the Morris water maze. Psychopharmacology. 2006;188(4):619–628.

[97] Tan KR, Rudolph U, Lüscher C. Hooked on benzodiazepines: GABAA receptor subtypes and addiction. Trends Neurosci. 2011;34(4):188–197. DOI: 10.1016/j.tins.2011.01.004.

[98] Rudolph U, Knoflach F. Beyond classical benzodiazepines: novel therapeutic potential of GABAA receptor subtypes. Nat Rev Drug Discov. 2011;10(9):685–697. DOI: 10.1038/nrd3502.

[99] Fritschy JM, Möhler H. GABAA-receptor heterogeneity in the adult rat brain: differential regional and cellular distribution of seven major subunits. J Comp Neurol. 1995;359(1):154–194.

[100] Sieghart W. Unraveling the function of GABA(A) receptor subtypes. Trends Pharmacol Sci. 2000;21(11):411–413.

[101] McKernan RM, Whiting PJ. Which GABAA-receptor subtypes really occur in the brain? Trends Neurosci. 1996;19(4):139–143.

[102] Holm KJ, Goa KL. Zolpidem: an update of its pharmacology, therapeutic efficacy and tolerability in the treatment of insomnia. Drugs. 2000;59(4):865–889.

[103] Biggio G, Concas A, Corda MG, Serra M. Enhancement of GABAergic transmission by zolpidem, an imidazopyridine with preferential affinity for type I benzodiazepine receptors. Eur J Pharmacol. 1989;161(2–3):173–180.

[104] Sanna E, Busonero F, Talani G, Carta M, Massa F, Peis M, et al. Comparison of the effects of zaleplon, zolpidem, and triazolam at various GABAA receptor subtypes. Eur J Pharmacol. 2002;451(2):103–110.

[105] Quirk K, Blurton P, Fletcher S, Leeson P, Tang F, Mellilo D, et al. [3H]L-655,708, a novel ligand selective for the benzodiazepine site of GABAA receptors which contain the $\alpha 5$ subunit. Neuropharmacology. 1996;35(9–10):1331–1335.

[106] Sternfeld F, Carling RW, Jelley RA, Ladduwahetty T, Merchant KJ, Moore KW, et al. Selective, orally active gamma-aminobutyric acid A $\alpha 5$ receptor inverse agonists as cognition enhancers. J Med Chem. 2004;47(9):2176–2179.

[107] Samardzic J, Puskas L, Obradovic M, Lazic-Puskas D, Obradovic D. Antidepressant effects of an inverse agonist selective for $\alpha 5$ GABA-A receptors in the rat forced swim test. Acta Vet (Beograd). 2014;64(1):52–60. DOI: 10.2478/acve-2014-0006.

[108] Dawson GR, Maubach KA, Collinson N, Cobain M, Everitt BJ, MacLeod AM, et al. An inverse agonist selective for $\alpha 5$ subunit-containing GABAA receptors enhances cognition. J Pharmacol Exp Ther. 2006;316(3):1335–1345.

[109] Ballard TM, Knoflach F, Prinssen E, Borroni E, Vivian JA, Basile J, et al. RO4938581, a novel cognitive enhancer acting at GABAA alpha5 subunit-containing receptors. Psychopharmacology (Berl). 2009;202(1–3):207–223. DOI: 10.1007/s00213-008-1357-7.

[110] Nutt DJ, Besson M, Wilson SJ, Dawson GR, Lingford-Hughes AR. Blockade of alcohol's amnestic activity in humans by an alpha5 subtype benzodiazepine receptor inverse agonist.Neuropharmacology. 2007;53(7):810–820.

[111] Thachil AF, Mohan R, Bhugra D. The evidence base of complementary and alternative therapies in depression. J Affect Disord. 2007;97(1–3):23–35.

[112] Wasowski C, Marder M. Flavonoids as GABAA receptor ligands: the whole story? J Exp Pharmacol. 2012;4:9–24. DOI: 10.2147/JEP.S23105.

[113] Luk KC, Stern L, Weigele M, O'Brien RA, Spirst N. Isolation and identification of "diazepam-like" compounds from bovine urine. J Nat Prod. 1983;46(6):852–861.

[114] Wolfman C, Viola H, Paladini AC, Dajas F, Medina JH. Possible anxiolytic effects of chrysin, a central benzodiazepine receptor ligand isolated from Passiflora coerulea. Pharmacol Biochem Behav. 1994;47(1):1–4.

[115] Vissiennona C, Nieber K, Kelber O, Butterweck V. Route of administration determines the anxiolytic activity of the flavonols kaempferol, quercetin and myricetin – are they prodrugs? J Nutr Biochem. 2012;23(7):733–40. DOI: 10.1016/j.jnutbio.2011.03.017.

[116] Dekermendjian K, Kahnberg P, Witt MR, Sterner O, Nielsen M, Liljefors T. Structure–activity relationships and molecular modelling analysis of flavonoids binding to the benzodiazepine site of the rat brain GABAA receptor complex. J Med Chem. 1999;42(21):4343–4350.

[117] Avallone R, Zanoli P, Puia G, Kleinschnitz M, Schreier P, Baraldi M. Pharmacological profile of apigenin, a flavonoid isolated from Matricaria chamomilla. Biochem Pharmacol. 2000;59(11):1387–1394.

[118] Nilsson J, Sterner O. Modulation of GABA(A) receptors by natural products and the development of novel synthetic ligands for the benzodiazepine binding site. Curr Drug Targets. 2011;12(11):1674–1688.

[119] Sasaki K, Hatta S, Haga M, Ohshika H. Effects of bilobalide on gamma-aminobutyric acid levels and glutamic acid decarboxylase in mouse brain. Eur J Pharmacol. 1999;367(2–3):165–173.

[120] Bone K, Mills S. Principles and Practice of Phytotherapy. 2nd ed. Edinburgh: Churchill Livingstone; 2013. p. 601–602.

[121] Johnston GA, Hanrahan JR, Chebib M, Duke RK, Mewett KN. Modulation of ionotropic GABA receptors by natural products of plant origin. Adv Pharmacol. 2006;54:285–316.

Chapter 3

The Integration of Psychodynamic Theories and Biological Aspects in the Development of Anxiety and Anxiety Disorders

Rudolf Gregurek and Rudolf Gregurek

Additional information is available at the end of the chapter

http://dx.doi.org/10.5772/65669

Abstract

Anxiety is a normal human reaction to stressful and threatening events in their surroundings, but if it does not correlate with inducible stimulus with respect to intensity and duration and if it permanently impairs a person's ability to function normally, then we are dealing with pathological anxiety, that is to say, a symptom of one of the anxiety disorders classified in 9th revision of the International Classification of Diseases (ICD-10), or in American Psychiatric Association, DSM-IV classification. We may consider the aetiology of anxiety from psychodynamic, biological and neuroscientific aspect. Finally, certain genes have been located, the variability of which in the expression of "visceral brain" neurons modulates remembrance of fear and somatic reactions to anxiety. These genes also represent the potential focal points for future pharmacotherapeutic solutions for the treatment of anxiety and anxiety-connected psychic disorders.

Keywords: anxiety, aetiology, psychodynamics, neuroscience, biological aspects, neuroscientific aspects

1. Introduction

Throughout evolution each species has been armed with certain defence mechanisms to help the organisms cope with internal or external threats. These defence mechanisms are highly individual from one species to another.

The term anxiety can also be used to describe normal reactions to danger, which can then be differentiated from fear because anxiety is much weaker and lasts longer than fear; however, anxiety can also be a pathological state of mind, because constant fear of unknown, undefined or unreal comes from inside the person.

Fear is the best activator of stress response because it has a vast array of peripheral and neuroendocrine changes which help us survive. Fear also impacts the brain, altering the way emotional and cognitive systems process information [1]. Even though all these mechanisms help us when needed, they are a later threat because of chronic activation which when triggered can cause post-traumatic stress disorder (PTSD) and even some physical symptoms [2].

Fear is an emotion which has a central role in evolution and natural selection. It is also one of few emotions which we share even with lower mammals. Because of this connection, it is easy to research fear and its effect on animals and compare the result with human behaviour.

Psychoanalysis provided a highly influential theory for the development of anxiety also showing the way anxiety was maintained. Psychodynamic notions of anxiety have since lost favour to the more parsimonious and empirically grounded behavioural theories. These theories have a strong ground in treatment of anxiety. On the other hand, the main problem of behavioural theories is that they cannot be applied to explain all cases of anxiety.

Every person throughout their lifespan has experienced the feeling of anxiety. Fear and anxiety have become such strong synonyms that people are having a hard time differentiating between the two. According to Freud, anxiety is the fear of unknown, when a person cannot describe what they are afraid of. When a person is able to describe exactly what they are afraid of, the term fear is used.

Emotions affect our behaviour; every day we experience a lot of emotions most of which do not have a great effect on our behaviour. But when the intensity of emotions shifts up, they can have a huge impact on our behaviour [3].

The primary emotions are anger, fear, pleasure, sadness and disgust [4]. Emotions help us find balance in our surrounding, and they also help with adapting our behaviour to the situation we are faced with. They can be divided into two groups: positive and negative. Negative emotions express an attempt or intention to exclude. Strengthening one's own position at the expense of others. Keeping bad stuff away, destroying what is perceived as a threat. Negative emotions are fueled by an underlying fear of the unknown, a fear of the actions of others, and a need to control them or stop them to avoid being harmed.

Positive emotions express an attempt or an intention to include. Taking the whole into consideration. Working on learning more viewpoints, interacting more with others, enjoying making things better. Positive emotions are fueled by an underlying desire for enjoyment and unity.

Anxiety is a sign of life and experience, and it does not always have to be a part of an illness. Anxiety is a psychological, physiological and behavioural state induced in animals and humans by a threat to well-being or survival, either actual or potential. In biology, the term anxiety is associated with awareness and stimuli of the conscious on danger. When a living organism faces danger, it reacts in only two ways: fight and flight. When a human faces danger, the first sign is anxiety. Anxiety can stimulate a human being to fight, retreat, exhibit hyper-

http://dx.doi.org/10.5772/65669

activity and awareness. If anxiety activates awareness and readiness, then the person can overcome all the difficulties that they are faced with, in order to achieve their aim. This type of anxiety is called normal, and we benefit from normal anxiety. It helps us become aware and ready for all the problems we have to deal with, and it also awakes a sense of satisfaction and joy when the task is complete [5].

To understand anxiety on the biological level, it is necessary to be familiar with the physiology and anatomy of the brain and neurotransmitters. Pharmacological research shows us that in anxiety there is hyperactivity of the noradrenergic nuclei, locus coeruleus and noradrenergic pathways. Also it was noticed that lower levels of serotonin and gamma-hydroxybutyric acid (GABA) neurons cause anxiety disorders. Technical breakthroughs help us pinpoint certain regions and structures of the central nervous system which are responsible for conditioning and integrating the felling of fear and producing an adequate response. These neuroscientific insights are based on the Pavlovian (classical) conditioning. Modern neuroimaging techniques have located Papez circuit, limbic system and the complex of amygdala nuclei, which is the primary structure responsible for the processes of learning and fear conditioning [6].

2. Psychodynamic aspects of anxiety

Glen Gabbard provides a modern understanding of the psychodynamic aspects of anxiety disorder [7]. One can only admire the insights he derives from Freud's seminal work on the subject, *Inhibitions, Symptoms and Anxiety* [8]. Many of Freud's ideas are now being proven correct by neuroscientists. It has been proven that there is a memory system for anxiety responses in the amygdala that processes stimuli without any reference to conscious memory. Gabbard points out that panic attacks may not just happen without any reason but are more likely to happen because of stress memory activation which is specific to each patient. He also discusses panic disorders which find cause in childhood. These can be compared to scientific finding on animals which were separated in early life from their parents showing later in life as an anxiety prone adult animal [7].

The creator of the psychoanalytic paradigm is Sigmund Freud. The psychoanalytic model, which is based on Freud's works and later expanded by his followers and other psychoanalysts of the modern times, now includes the ideas of all those who reviewed and tried to change his mind. The model is based on the assumption that human behaviour is determined by intra-psychical impulses, desires, motives and conflicts.

For Freud, there are three stages in the development of the concept of anxiety. The first period (works between 1893 and 1895) is in connection with the neurosis of fear and its relationship with sex life [9]. In the second period (between 1909 and 1917), Freud elaborated the relationship between anxiety and repressed libido [10–13], and in the third period (between 1926 and 1932), he discusses the relationship of anxiety with the mental apparatus [8, 14, 15].

In his major work, "Inhibition, symptom and anxiety", Freud gave his final views on the theory of anxiety [8].

The alarm signal is a form of anxiety that occurs at provoking some old situations of danger and is a manifestation of the ego, which is used in order to start defensive measures against drives coming from the id or its representations. Defence mechanisms of the ego, no matter how much unskilled they are, condition the symbolic activity of a similar opinion.

Freud elaborated the relationship between anxiety, pain and grief for the object. Pain is a personal reaction to the loss of an object, while anxiety is a reaction to the danger that implies that loss and, through shifting, a reaction to the threat of the loss. The loss of the object causes the pain to penetrate insurmountable amount of excitation to the ego, which is experiencing anxiety because of fear of helplessness. In order to prevent this occurrence of pain and fear of helplessness, anxiety signal precedes the disaster and warns ego to employ defensive measures that will be able to cope with danger [14].

To sum up, this theory is based on the concept of danger and "dangerous situations" and is based on two main pillars: (1) anxiety occurs as a response signal for the purpose of preparing a person to a dangerous situation and (2) the ego is the centre of anxiety and sometimes can even be the cause, whether it is repeated anxiety for their own account (eroticized anxiety) or as a signal of impending danger instinctively. Functional anxiety is determined by two aspects:

1. historical aspect, since the anxiety as a signal represents a repetition of infantile anxiety experiences, which it plays, creating at the same time some protection from the return of the repressed;
2. symbolic aspect, since it is a function of anxiety at the same time and symbolic because it represents in itself symbolically a dangerous situation.

As a universal reaction of the human being, anxiety frequently occurs as a situational conditioned disorder, and only further observation usually allows us to distinguish normal, neurotic and psychotic anxiety.

2.1. Object relations theory

Beside Freud's theory, in the later development of psychoanalysis, a second significant theory arises—object relations theory. Without this theory, we would not be able to explain the necessity to communicate with other individuals [16–18].

Object relations theory stresses the importance of the earliest interactions with other people from our surrounding as a building stone in the construction of the id, ego and superego. More specifically, from birth onwards, our relations with others are highly influenced by strong emotions which form specific memories and integrate us in the society by our behaviour. Positive experience helps us form ourselves as mature individuals, and the same goes for negative experience. Each experience is individually processed and stored in our memory [19].

Simply put, we can say that we are born with a certain capacity of perception, memory and establishing a representation of our perception and to gradually develop symbolic thought, or the capacity of abstract thinking and intelligence. Let us imagine that the ego is a computer that absorbs, stores and integrates information and learns how to sort and classify the specific pattern of priorities. It also has to distinguish the important from the unimportant, good from

http://dx.doi.org/10.5772/65669

bad. Gradually, we learn and differentiate what is inside from what is outside, and so, overtime, our inner world is built. The largest part of it remains in the unconscious memory, or preconscious area (reservoir of information that we do not use all the time, but can access), while only a small part goes into the realm of conscious. The rest is stored in an even deeper level, the dynamic unconscious or id [20].

3. Biological aspects of anxiety

Neurochemistry is the only way to understand and cure anxiety disorders. Pharmacological treatment is based on neurochemistry. Neurons communicate with each other with neurotransmitters; at least three neurotransmitters can be associated with anxiety. In recent studies, there is a sign that certain large neurotransmitters also play a role in anxiety and they are called neuropeptides. This neuropeptides are not only used in communication between neurons, but can also regulate them and cause hormonal misbalance. This is because of their chemical structure which biologically has a strong effect on the human body. Research on neuropeptides is still at its beginnings, so it will not be mentioned any more through this chapter because there is no clinical finding which can be used to help us. For now, we will base ourselves on gamma-hydroxybutyric acid (GABA). When this neurotransmitters balance is disrupted in the body, we can note a relationship with anxiety. Benzodiazepines, drugs that exert their effect through GABA receptors, are used to help people suffering from acute anxiety [21].

Serotonin is greatly important in anxiety. It is a monoamine neurotransmitter involved in controlling a wide range of behaviours by affecting the neural system. These behaviours include emotions which are connected with fear and anxiety. All serotonin neurons emerge from the raphe nuclei. The raphe nuclei have a part called median raphe which acts as a connection with the septo hippocampal system as well as the cortex. Because of these connections, it has an important role in emotional cognition [22].

Studies on animals have successfully proved the connection between different neurotransmitters and anxiety. These types of studies are called knockout studies, because a certain receptor in the animal genetic code is knocked out. Animals which have had their serotonin re-uptake transporters knocked out show in case studies abnormal response to fear and anxiety in a number of behavioural conflict tests. This confirms the role of this receptor in modulating anxiety [23].

On the other hand, animals which had their serotonin receptor knocked out showed an increased heart rate and anxiety in a large variety of tasks such as eating and locomotion. Foot shock on animals with the serotonin receptor removed showed longer freezing and increased tachycardia.

Together these data show the importance of serotonin in modulating the levels of fear and anxiety in our brain.

As stated above, the median raphe nuclei (MRN) provide the necessary input to the neural circuits within the brain causing fear and anxiety modulation. MRN lesion showed us that short-term fear is independent from MRN, while long-term fear depends on MRN [24].

In human research, it was found that people with short serotonin transport acquired faster fear than people with longer serotonin transport. All these findings are consistent with the findings in animal models.

However, the best proof of the function of the serotonergic system in fear and anxiety is the pharmacological evidence. Drugs that change the function of serotonin have beneficial effects on various forms of anxiety. The best pharmacological treatment of anxiety is serotonin re-uptake inhibitors which allow greater levels of serotonin to accumulate and in that way help in treatment of anxiety [22].

Noradrenalin is also an important neurotransmitter in anxiety. Neurons which carry noradrenalin rise from the locus coereuleus (LC), and these are also a centre associated with warnings or alarms. The LC secretes directly into the brain causing immediate response. Increased levels of noradrenalin cause higher levels of anxiety. In pharmacology, noradrenalin blockers lower the levels of noradrenalin and do the same to a patient as do serotonin re-uptake blockers. Among adults, agents that alter noradrenergic functioning are powerful anxiolytics. Similarly, agents, such as yohimbine, that increase firing of the locus coeruleus are potent anxiogenic compounds [25].

3.1. The visceral brain

At the beginning of the twentieth-century, scientists have identified the hypothalamus as a key structure in the control of the autonomic nervous system [26]. Based on these assumptions, there was a so-called hypothalamic theory of emotion, which contained three main strands: first—the hypothalamus in a way valued events in our environment; second—the expression of emotional response is proportional to the outbreak of the pulses from the hypothalamus in the brain stem; and the third—hypothalamic projections into the cortex allow conscious perception of emotion [27, 28]. American neurologist James Papez in 1937 proposed that the circuit connecting the hypothalamus to the limbic lobe was the basis for emotional experiences. Papez circuit (or medial limbic circuit) is a neural circuit for the control of emotional expression [29]. Papez theory was later reconceptualized and expanded by an American neuroscientist Paul D. MacLean [30]; MacLean redefined the circuit as the "visceral brain", which consisted of the limbic lobe and its major connections in the forebrain—hypothalamus, amygdala and septum. Overtime, the concept of a forebrain circuit for the control of emotional expression has been modified to include the prefrontal cortex. The amygdale, complex nuclei (central, medial and cortical basolateral, which is still divided on the lateral, basal and accessory basal nucleus) located in the medial temporal lobe was also part of MacLean's theory of the limbic system, but not of any importance, until the 1956 publication by the British psychologist Lawrence Weiskrantz. In this work, he tried to explain the causes of clinical manifestations of the Kluver-Bucy syndrome (a set of behavioural disorders due to the bilateral damage to the temporal lobes usually after temporal lobotomy).

http://dx.doi.org/10.5772/65669

Weiskrantz suggested that the main cause of such a clinical picture is just the damaged or removed amygdala. For years after its publication, numerous studies have focused on the amygdale function and its relationship with other parts of the central nervous system; however, greater progress in understanding the role of the functions of the amygdala and its link with emotions, including fear, has been in the 70s and 80s, when the scientists returned Pavlov method of classical conditioning [31].

3.2. The amygdala in fear conditioning process

Evolutionary man has developed a number of coping mechanisms, some of which are based on the ability to anticipate and avoid dangerous, potentially life-threatening situations. Neural circuits involved in the processes of memory and processing emotions play a key role in these mechanisms; each new stimulus is incorporated and based on the nature of that stimulus and appropriate behavioural, psychological and somatic reaction that, in addition to character, must match the intensity of the resulting stimulus. The scientists used this knowledge in research of pathways related to emotions, while the main tool used was the simple method of classical conditioning, similar to that used by Pavlov in his experiments. In it he explored the mechanisms by which animals predict a dangerous or a pleasant event. In these studies, emotionally related stimulus (unconditioned stimulus), which was the unpleasant, threatening, associated with fear, or rewarding, and reward-related event, preceded by an emotionally neutral (s conditioning) stimulus, mainly sound or light pulse [32–35].

After a period of learning (conditioning), animals are using these conditioned stimuli to predict the magnitude and time of occurrence of the desired or undesired events. In further analyses of conditioned fear, the researchers were able to map the entire neural path from the input of sensory stimuli to output behavioural responses [36–38]. Studies have shown that just amygdala, a complex structure of the temporal lobes, plays a key role in the coordination of behavioural and psychosomatic reactions associated with fear [37–42]. When sound (neutral stimulus) arrives to organ of Corti to signal danger, impulse to the amygdala comes from two ways. The first is through the thalamus, through nucleus corpora geniculate medial, responsible for fast processing of sound information that allows quick preparation of the body to potentially dangerous and threatening event before sound region of the cerebral cortex of the temporal lobe processes the audio information and discern whether it is a real danger or "false alarm" [40]. The second path, which leads from the sound region temporal cortex to the amygdala, is slower, but also more complex because it integrates functions of the higher centres of the central nervous system. Electrophysiological studies have shown that conditioned stimuli that preceded those unfavourable fear conditioning, then the fear associated, strengthens synaptic connections between the auditory thalamus and lateral nuclei of the amygdala. The stimulus is passed into the central amygdala nuclei that project axons to different regions of the hypothalamus, activating the sympathetic system and inducing the secretion of stress hormones such as CRH (corticotropin-releasing hormone). The secretion of CRH in the periventricular nuclei of the hypothalamus results in an increased secretion of adrenocorticotropic hormone (ACTH) in adenohypophysis and consequently increases the secretion of glucocorticoids from the adrenal cortex, by which the body leads to a catabolic state. The central nucleus of the amygdala plays an important role in emotional behavior. Nociceptive inputs through the spino-parabrachio-amygdala pathway probably contribute to pain-induced changes in affective behavior, and the projections of the amygdala to the PAG-RVM (rostroventromedial medulla) circuitry may be involved in mediating the influence of emotions on pain. Stressful situations like physical exercise, exposure to extreme temperatures, fight, fear and pain may induce a decrease in pain sensitivity. However, if these mechanisms are deregulated, it creates a strong etiological factor backing the development of anxiety disorders [43].

4. Case study: example of connection psychodynamic and neuroscientific approach to treatment of anxiety

In recent years, there are numerous mental disorders as the consequences of extreme stress; they are the official classifications of anxiety and PTSD. Post-traumatic stress disorder (PTSD) develops in some people after exposure to an extreme stress or traumatic event. It is characterized by three distinct types of symptoms consisting of re-experiencing of the event, avoidance of reminders of the event and hyper arousal, which continue for a substantial period of time. These symptoms indicate dysfunction in acquisition and extinction of fear conditioning, and the hippocampus has been the most consistently implicated brain region.

Our clinical experience has shown that hyper arousal symptoms and especially those associated with impulsive aggressiveness, such as hyperirritability, hyper excitability, hypersensitivity, aggressive acting outs, outbursts of anger, present one of the dominant disturbances for chronic PTSD patients. This kind of symptoms can be understood in the perspective of loss of the affective modulation, which can explain the PTSD patients' lack of capacity to use effect states as cues to attend to incoming information. The elevated—arousal—state is likely to precipitate flight-or-fight reactions in these patients, so they often go immediately from stimulus to response without psychologically assessing the meaning of the event. This makes them prone to freeze or, alternatively, to overreact in response to minor provocations [44].

The aim of our study was to assess possible changes in brain activation in PTSD patients after 2 years of intensive psychotherapy in form of closed-door group analysis in a heterogeneous group specially designed in our clinic for treatment of PTSD patients.

4.1. Method and participants

Participants in our study were selected from all PTSD patients treated at the Clinic for Psychological Medicine, University of Zagreb Medical School, in the period from April to July 2001. We selected 25 participants from the total of 82 patients treated in the day hospital unit at the clinic over the period. All the participants in the study, as the most of our patients, were Croatian war (1991–1995) veterans who actively participated in the combat and were exposed to multiple traumatic experiences such as witnessing the death of the fellow soldiers and sudden air-rides or artillery attacks. The patients who experienced other kinds of traumatic experiences such as prisoners of war and concentration camp detainees were not included in this study (because they have difficult symptoms, and they need psychopharmacological therapy, and other reason was ethical because we do not want to retraumatize these vulnerable patients). They all signed informed consent to participate in the study.

The group evaluation was performed over 2-week group psychotherapy, as described above, encompassing four group sessions and consisted of monitoring of patients' symptom manifestation. The same co-therapist couple performed the selection for all the participants in the study. Individual evaluation was organized in the form of Structured Clinical Interview for DSM-IV (SCID-CV) [45]. The interview was held during the first week of patients' treatment at the Clinic and was conducted by the same therapist for all the participants. All therapists conducting the selection were clinicians with at last 3 years of experience in working with psych traumatized patients. Group therapists were not informed of the evaluation performed by the individual therapist, conversely.

All patients selected for the investigation had chronic, combat-related PTSD with severe hyper arousal symptoms, impulsive aggressiveness and no other Axis-I and/or Axis-II diagnosis in accordance with DSM-IV criteria (American Psychiatric Association, 1994) [46]. Severity of hyper arousal symptoms was assessed on the basis of anamnesis and heteroanamnesis and also on the basis of symptoms they manifested in group therapy and individual psychiatric interview. Only the patients who continuously over the two-week observation presented all hyper arousal symptoms as defined in DSM-IV, and impulsive aggressiveness symptoms were selected for the study. Impulsive aggressiveness symptoms were defined as: hyperirritability, aggressive acting outs and outbursts of anger. Only patients who manifested these symptoms in both individual interviews and all four group sessions over the observation period, and who had positive anamnesis and heteroanamnesis for hyper arousal symptoms in the month preceding the hospitalization, were included in the study. This selection design had to be implemented because no psychometric instrument for assessing hyper arousal and/or impulsive aggressiveness symptoms as such, qualitatively or preferably quantitatively, was available in Croatia.

The patients selected for the study received no other than benzodiazepine pharmacotherapy for at least 6 months preceding the study. Also, 4 weeks before the study entry the patients received no psychotropic medication at all. After the psychiatric evaluation, neurological examination was conducted for all participants to exclude those with possible neurological comorbid conditions, head trauma or loss of consciousness in the year preceding the study.

All patients included in the study were male, right-handed, of similar age (mean = 42 years, SD = 2) and had no history of alcohol or drug abuse or dependence.

Single photon emission computed tomography (SPECT) using Tc^{99}-ECD functional brain imaging was organized in the week after the two-week observation period immediately after the first group session in that week to establish possible alterations in regional cerebral blood flow (rCBF). The procedure was organized in the following way: patients who manifested at least one and no more than three aggressive acting outs during the pre-SPECT session were asked to remember the specific situation in the group that provoked the last aggressive acting out. The ultimate of no more than three aggressive acting outs during the designated session was a preset criteria to enable remembering of the specific situation that provoked the last acting out, but was not used because none of our subjects actually manifested more than three aggressive acting outs during the pre-SPECT session. The subjects were asked to try to remember the exact situation in the group that provoked their aggressive acting out and to describe it, and they were included in the further procedure. Once again, all of our subjects fulfilled these criteria and all declared very distressed when remembering the situation. Following this, subjects were introduced into SPECT procedure, and the scans were obtained in 50–60 min after the end of the session. Once again, immediately before the SPECT brain scan was made, each subject was asked to remember the situation that provoked his last aggressive acting out in the pre-SPECT session.

SPECT scans were started 20 min after the administration of 740 MBq of 99mTc-ECD in resting state ("white noise" environment, without any auditory or visual combat stimuli provocation). Measurements were made with IRIX triple-headed rotating scintillation gamma camera fitted with high-resolution collimators, using a 360° circular orbit and step and shoot mode (20-s image each 3°), so that the acquisition lasted 40 min. Within-subject correlated slices were obtained.

Following regions of interest (ROI) were considered: temporal, frontal, orbitofrontal and occipital cortices, cingulate gyrus, amygdala, thalamus, basal ganglia (caudate, putamen, ventral basal ganglia) and cerebellum. The irregular ROIs were outlined in the right hemisphere and mirrored in the left. The same investigator, blinded for all clinical data, placed ROIs. For each ROI, mean counts per pixel in two consecutive slices were averaged to minimalize partial volume effects. In each patient, studies were normalized to the mean whole-brain uptake, so relative hypo/hyper perfusion was established for each ROI. The percentage of asymmetry between two homologous regions was calculated as 200× (right-left/right-left), where right and left indicate mean average counts per pixel in the ROIs of the respective hemisphere. The negative values indicate thus lower perfusion in the right hemisphere. Percentages of interhemispheric asymmetry between homologous brain regions were used to identify abnormalities.

4.2. Therapy

Patients were subdued to a group psychotherapy used in the form of closed-door group analysis with once a week 90-min session rate, focusing on patients' interpersonal and social relations and communication, specially designed in our clinic for treatment of PTSD patients. The group consisted of 10 members: six men (Croatian war veterans) and four women (partners of the war veterans, who were not related to the men in the group). During the psychotherapy period, the patients consumed no pharmacotherapy, except anxiolytic (benzodiazepine) therapy over the first 3 weeks of the therapy in order to alleviate acute symptoms and enable the psychotherapeutic process.

People whose reactions to early traumas have become integrated into the totality of their personalities are apt to repeat aspects of the trauma and defensive reactions to it in their relationships with other people. In heterogeneous group, transference-oriented trauma

http://dx.doi.org/10.5772/65669

groups, the social expressions of those trauma-related affects and cognitive schemata are inevitably expressed on a social level and became readily observable.

Long-term heterogeneous groups can be particularly helpful to people with previous individual therapy, where they had an opportunity to develop a language in which they can identify their feelings and a basic curiosity about how they may contribute to their problems in interpersonal relationships themselves. Patients "fitting" for the group analysis should have the ability to observe their own problems, the ability to react emotionally as well as a certain capacity to save somebody else's emotions and the capacity for insight [47].

The therapeutic environment should be created in that way to enable individuals who are similar in some way to understand and feel with others, but in the same time different enough so they could watch and help other group members from different perspectives and positions. The best group composition is if the members are similar in terms of ego development and different in terms of interpersonal style. Groups that focus on the relationships between the members allow new losses to be experienced, as object loses with concomitant grief and sadness, rather than with narcissistic injuries accompanied by feelings of helplessness, numbing and vengeful rage. As heterogeneous trauma groups mature, they gradually make that transition. When that happens, the group members start getting the full benefit of the strengths that traumatized persons have developed to survive catastrophic trauma.

4.3. Results

Based on the studies of normal subjects, it is well documented that the normal cerebral blood distribution is bilaterally symmetric, with percentages of interhemispheric asymmetry never exceeding 12% for any brain region [48, 49]. In our study, SPECT brain scans at the beginning of the psychotherapeutic treatment, with two-week period between, showed unilateral (dominant brain hemisphere) 20% or more increase in the regional blood flow (rCBF) in the projection area of ventral basal ganglia (VBG).

After 2 years of group psychotherapy, PTSD patient's symptoms diminished, and they regained adequate impulse control. Control SPECT scans, of the same group of subjects, were performed in two separate occasions with 3-week period in-between. No differences from normal pattern were found. Because of that reason, we do not need control group, because subjects acted as their own control (before and after group psychotherapy).

5. Conclusion: the integration knowledge of psychodynamics and biological aspects in treatment of anxiety disorder

In our study, we found unilaterally increased perfusion in the projection area of ventral basal ganglia (VBG) in the dominant brain hemisphere, in our PTSD subjects with severe hyper arousal symptoms and impulsive aggressiveness, and after 2 year of group psychotherapy, our patients show no differences from normal pattern. Possible conclusion is that SPECT findings are associated with the symptoms of PTSD. They receded together with the clinical manifestation of symptoms, so it could be used as neurobiological indicator in order to make psychotherapeutic intervention focused on symptoms and clinical remission yet neurobiologically verified, as in our study.

Trauma is biochemically encoded into the brain in a variety of ways, including changes in the availability and effects of neurotransmitters and neuromodulators.

Neuroanatomical encoding occurs through changes in structures like the hippocampus, through coordination and integration of neural network functioning. The hyper arousal symptoms and persistent re-experiencing of the traumatic event suggest abnormalities in emotion and memory regulation implicating thus limbic brain regions as possibly associated with the disorder.

As scanning techniques become more precise and the hardware more affordable, they will without doubt become incorporated into treatment of PTSD and other disorders. As a part of initial assessment, they could help therapists pinpoint areas of neural activation and inhibition.

Treatment planning will come to include specific psychotherapeutic and pharmacological intervention to enhance growth and integration of affected networks.

Regular scans during the course of therapy may someday replace psychological tests as ways of fine-tuning the therapeutic process and measuring treatment success. The psychiatrist must be thought to think about the whole person first, to appreciate that each one is interesting and unique, not simply a composite of symptoms that are used to make an ICD or DSM diagnosis and provide treatment according to a standard algorithm.

It is possible to conclude that SPECT findings are associated with the symptoms of PTSD.

They receded together with the clinical manifestation of symptoms, so it could be used as neurobiological indicator in order to make psychotherapeutical intervention focused on symptoms and clinical remission yet neurobiologically verified, like as in Seedat study [50]. Functional brain imaging has opened a window to the living human brain in the acts of motor tasks, experiencing symptoms, imagining a feared situation or telling a lie.

An examination of areas activated during different tasks has enhanced our understanding of which neural networks participate in various functions [51].

Recent neurobiological research has confirmed Freud's original observation that there are essentially two forms of anxiety: one largely determined by psychological issues and another driven by autonomous biological factors outside the realm of psychological content. Research of locus coeruleus has been productive in defining the biological dimensions of anxiety. Whereas Freud viewed the ego as the psychological seat of anxiety, modern neurobiological researchers have identified the locus coruleous as the biological seat of anxiety. Neural pathways enter this nucleolus from every level of the central nervous system, and efferent pathways lead to all the major psychological systems involved in anxiety.

The existence of a neural mechanism for anxiety does not, however, preclude the usefulness of psychotherapeutic techniques. Effective psychotherapy most likely results in long-term

http://dx.doi.org/10.5772/65669

structural and functional changes in the brain that are related to changes in the expression of genes.

Contemporary approaches to the treatment of mental disorders are not distinguished from older ones by emphasis on drugs as by an emphasis of effective treatment. Psychotherapy is fundamentally a learning process for its patients and as such is a way to rewire the brain. In this sense, psychotherapy ultimately uses biological mechanisms to treat mental illness. This does not mean, however, that psychotherapy involves learning, while drug therapy involves something else, like correction of genetically dictated chemical imbalance. Even if chemical imbalance was to account for mental disorders, the imbalance could result purely from environmental factors, like some intensely stressful experience, or from environmental events that trigger and amplify a genetic predisposition. The therapist's job, whether using drugs or not, is to restore mental well-being [52].

At the centre of psychotherapy is an understanding of mutual interweaving of nature and education, successful and unsuccessful development and its impact on the healthy functioning. When psychotherapy results in a reduction in severity of symptoms, the brain is, in some ways, changed [52].

Fundamental premise was "that any form of psychotherapy is successful to the degree to which it enhances positive experiential change and underlying neural networks, growth and integration". All forms of psychotherapy are successful to the extent they enhance change in the relevant neural circuits. The brain is an organ of adaptation and built by experience during development and rebuilt during psychotherapy. At the neural level, integration and communication of neurons are associated with feelings, cognition and behaviour. On the psychological level, integration is marked with the ability to experience important life situation to be dismissed with the inclusion of minimum defence.

From the perspective of neuroscience, psychotherapy can be understood as a special form of enriched environment created to facilitate the growth of neurons and the integration of neural connections. The therapeutic environment is, in fact, an environment tailored individually, according to the symptoms and the patient's needs [52].

Historically speaking, before the Freud's treatment of psychiatric disease was the area of neurology. Even Freud, who was a neurologist, believed that most psychiatric disorders have an organic substrate and anticipated application medication. Freud wrote in 1914: "We must be aware of the fact that our provisional ideas in psychology probably one day be based on organic substructures". In the article, "The draft of psychoanalysis" Freud says: "Maybe in the future we learn to direct, through chemical substances, influence the quantity of energy and its distribution in the mental apparatus" [53].

In the years that followed, the more emphasized the difference in the two approaches became, but its peak is experienced in the 50s when the current psychodynamic psychiatry equated with intellectual and better educated grouping of pharmacologically oriented psychiatrists. Neuroleptics and antidepressants, which were discovered in the late fifties, have not been as effective in the treatment, but research has increasingly intensified, and the drugs have become more efficient, which resulted in the DSM-III-R classification of diseases that did not include psychodynamic knowledge in diagnosis mental illnesses [54].

Our neural web is determined by two types of effects: first one is genetics: this part is individual from person to person; and the second one is external factors: it includes the environment which moulds our genetics. This process of moulding genetics is to actually activate or inactivate certain genes that are not used causing parts of the neural network to stop working due to the inactivation. Moulding is dependent on multiple life style factors, like diet and exercise [54].

Synaptic connections and communications do not go away. Psychotherapy works in such a way that it uses the brains plasticity to cause forming of new synaptic connections which will be stronger than the earlier ones and in cause be used more. Psychotherapy can also inhibit the usage of undesired synaptic connections. Of course, for Stahl, this is ideally achieved through the synergy with drugs and psychotherapy. He also states that medication can help the facilitation process and enhance the health situation [55].

All this research of learning and the way we memorize does not just help us understand psychotherapy, but it also gives us a new perspective of anxiety disorders involving the inability to control fear, obsessions, compulsions and delusions. To end this paper, it should be stated that the greatest leap forward would be a better collaboration between neurobiologists and psychotherapists in trying to find links between neurobiology and psychotherapeutic processes [56].

Author details

Rudolf Gregurek[1*] and Rudolf Gregurek[2]

*Address all correspondence to: rudolf.gregurek@mef.hr

1 School of Medicine Zagreb, University of Zagreb, Zagreb, Croatia

2 School of Medicine Rijeka, University of Rijeka, Rijeka, Croatia

References

[1] LaBar KS, LeDoux JE. Coping with danger: the neural basis of defensive behaviors and fearful feelings. In BS McEwen (ed.) Handbook of physiology, Section 7: the endocrine system, vol. IV: coping with the environment: neural and endocrine mechanisms, pp. 139–154. New York: Oxford University Press, 2001.

[2] Sapolsky RM, Romero LM, Munck AU. How do glucocorticoids influence stress responses? Integrating permissive, suppressive, stimulatory, and preparative actions. Endocrine Reviews, 2000;21:55–89.

[3] James W. The Physical Basis of Emotion. Psychological Review, 1894;1:516–529.

[4] Panskepp J. Affective neuroscience: the foundations of human and animal emotions. New York: Oxford University Press, 1998.

[5] Schore AN. A century after Freud's project: is a rapprochement between psychoanalysis and neurobiology at hand? Journal of the American Psychoanalytic Association, 1997;45:841–867.

[6] Barlow DH. Anxiety and its disorders: the nature and treatment of anxiety and panic. New York: The Guilford Press, 2002.

[7] Gabbard GO, Goodwin F. Clinical psychiatry in transition: integrating biological and psychosocial perspectives. In Dickstein LJ, Riba MB, Oldham JM (eds.) Review of psychiatry, vol. 15, pp. 527–548. Washington, DC: American Psychiatric Press, 1996.

[8] Freud S. Inhibitions, symptoms and anxiety.S. E., vol. 20, London: The Hogarthe Press, 1926.

[9] Freud S, Breuer J. (1893) Studies in hysteria. In Standard edition, vol. XVI, London: The Hogarth Press, 1966.

[10] Freud S. (1910) The future prospects of psychoanalytic therapy. Standard Edition, London: Hogarth Press, 1953.

[11] Freud S. (1912) Recommendations to physicians practicing psychoanalysis. Standard Edition, London: Hogarth Press, 1953.

[12] Freud S. (1912) The dynamics of transference. In J. Strachey (ed.) The Standard edition of the complete works of Sigmund Freud, vol. 12, pp. 97–108. London: Hogarth Press, 1958.

[13] Freud S. (1916) Fixation to traumas—the unconscious. In Standard edition, vol. XII. London: Hogarth Press, 1966.

[14] Freud S. Mourning and melancholia. Standard edition IV. London: Hogarth Press, 1966.

[15] Freud S. The interpretation of dreams. Standard edition IV. London: Hogarth Press, 1966.

[16] Gabbard GO. Psychodynamic psychiatry in clinical practice: the DSM-IV Edition. American Psychiatric Press, Washington, DC, 1994.

[17] Greenberg JR, Mitchell JR. Object relations in psychoanalytic theory. Cambridge, MA. Harvard University Press, 1983.

[18] Horner AJ. Psychoanalytic object relations therapy. Northvale, NJ: Jason Aronson, 1991.

[19] Klein M. Love, guilt and reparation. In The Writings of Melanie Klein, vol. 3, chapter 19, pp. 306–343. London: Hogarth Press, 1975, 1937.

[20] Klein M. Some theoretical conclusions regarding the emotional life of the infant. In M. Klein, P. Heimann, S. Isaacs, & J. Riviere (eds.) Developments in psychoanalysis, pp. 198–236. London: Hogarth, 1952.

[21] Zwanzger P, Rupprecht R. Selective Gabaergic treatment for panic, Investigations in experimental panic induction and panic disorder. Journal of Psychiatry Neuroscience, 2005;30(3):167–175.

[22] McKittrick CR, Magarinos AM, Blanchard DC, Blanchard RJ, McEwen BS, Sakai RR. Chronic social stress reduces dendritic arbors in CA3 of hippocampus and decreases binding to serotonin transporter sites. Synapse, 2000;36:85–94.

[23] Gross C, Santarelli L, Brunner D, Zhuang X, Hen R. Altered fear circuits in 5-HT-sub(1A) receptor KO mice. Biological Psychiatry, 2000;48(12):1157–1163.

[24] Melik E, Babar-Melik E, Oezguenen T, Binokay, S. Median raphe nucleus mediates forming long-term but not short-term contextual fear conditioning in rats. Behavioural Brain Research, 2000;112(1–2):145–150.

[25] Levine ES, Litto WJ, Jacobs BL. Activity of cat locus coeruleus noradrenergic neurons during the defense reaction. Brain Research, 1990;531:189–195.

[26] Karplus JP, Kreidl A. Gehirn und Sympathicus. VII: Uber beziehungen der hypothalamuszentren zu blutdruck und innerer sekretion. Pfluegers Arch. Gesamte Physiol. Menschen Tiere, 1927;215:667–670.

[27] Bard PA. Diencephalic mechanism for the expression of rage with special reference to the sympathetic nervous system. American Journal of Physiology Legacy Contents, 1928;84:490–515.

[28] Cannon WB. Bodily Changes in Pain, Hunger, Fear, and Rage, Appleton, New York, 1929.

[29] Snider RS, Maiti A. Cerebellar contributions to Papez circuit. Journal of Neuroscience Research, 1976;2(2):133–146.

[30] MacLean PD. Some psychiatric implications of physiological studies on frontotemporal portion of limbic system (visceral brain). Electroencephalography and Clinical Neurophysiology, 1952;4:407–418.

[31] Weiskrantz L. Behavioral changes associated with ablation of the amygdaloid complex in monkeys. Journal of Comparative and Physiological Psychology, 1956;49(4):381–391.

[32] Maren S. Neurobiology of Pavlovian fear conditioning. Annual Review of Neuroscience, 2001;24:897–931.

[33] Mountney C, Sillberg V, Kent P, Anisman H, Merali Z. The role of gastrin-releasing peptide on conditioned fear: differential cortical and amygdaloid responses in the rat. Psychopharmacology (Berl) 2006;189:287–296.

[34] Mountney C, Anisman H, Merali Z. Effects of gastrin-releasing peptide agonist and antagonist administered to the basolateral nucleus of the amygdala on conditioned fear in the rat. Psychopharmacology (Berl), 2008;200:51–58.

[35] Shah A, Jhawar SS, Goel A. Analysis of the anatomy of the Papez circuit and adjoining limbic system by fiber dissection techniques. Journal of Clinical Neuroscience, 2012;19(2):289–298.

[36] Buchel C, Dolan RJ. Classical fear conditioning in functional neuroimaging. Current Opinion in Neurobiology, 2000;10:219–223.

[37] Milad MR, Rauch SL, Pitman RK, Quirk GJ. Fear extinction in rats: implications for human brain imaging and anxiety disorders. Biological Psychology, 2006;73:61–71.

[38] Fendt M, Fanselow MS. The neuroanatomical and neurochemical basis of conditioned fear. Neuroscience and Biobehavioral Reviews, 1999;23:743–760.

[39] LeDoux J. Fear and the brain: where have we been, and where are we going? Biological Psychiatry, 1998;44:1229–1238.

[40] McNally JR. Mechanisms of exposure theraphy: how neurescience can improve psychological treatement for anxiety disorders, Clinical Psychology Review, 2007;6/7:750–759.

[41] Chaperon F, Fendt M, Kelly P, Lingenhoehl K, Mosbacher J, Olpe H, Schmid P, Sturchler C, McAllister K, van der Putten H, Gee C. Gastrin-releasing peptide signaling plays a limited and subtle role in amygdala physiology and aversive memory. PLoS One. 2012 7(4):e34963. doi: 10.1371/journal.pone.0034963. Epub 2012 Apr 11.PMID:22509372

[42] Maren S, Quirk GJ. Neuronal signalling of fear memory. Nature Reviews Neuroscience, 2004;5:844–852.

[43] Popov VI, Bocharova LS. Hibernation-induced structural changes insynaptic contacts between mossy fibres and hippocampal pyramidal neurons. Neuroscience, 1992;48:53–62.

[44] van der Kolk. The body keeps the score. In Horowitz MJ (ed.) Essential papers on Posttraumatic stress disorder. New York, London: New York University Press, 1999.

[45] First MB, Spitzer RL, Gibbon M, and Williams, JBW. Structured Clinical Interview for DSM-IV Axis I Disorders, Clinician Version (SCID-CV). Washington, D.C.: American Psychiatric Press, Inc., 1996.

[46] American Psychiatric Association. Diagnostic and statistical manual of mental disorders, 4th edition. Washington, DC: American Psychiatric Assotiation, 1994.

[47] Gregurek R. Countertransference problems in the treatment of a mixed group of war veterans and female partners of war veterans. Croatian Medical Journal, 1999;40(4):493–497.

[48] Juni JE, Waxman AD, Devous MD Sr, Tikofsky RS, Ichise M, Van Heertum RL, Holman BL, Carretta RF, Chen CC. Procedure guideline for brain perfusion SPECT using technetium-99m radiopharmaceuticals. Society of Nuclear Medicine. Journal of Nuclear Medicine, 1998; 39(5):923–926.

[49] Koyama M, Kawashima R, Ito H, Ono S, Sato K, Goto R, Kinomura S, Yoshioka S, Sato T, Fukuda H. SPECT imaging of normal subjects with technetium-99m-HMPAO and technetium-99m-ECD. Journal of Nuclear Medicine, 1997;38(4):587–592.

[50] Seedat S, Warwick J, van Heerden B et al. Single photon emission computed tomography in posttraumatic stress disorder before and after treatment with a selective serotonin reuptake inhibitor. Journal of Affective Disorders, 2004;80:45–53.

[51] Gregurek R. Posttraumatic stress disorder from neurophysiology to psychotherapy: a croatian experience. In Widera-Wysoczanska A, Kuczynska A (eds.) Interpersonnal trauma and its consequences in adulthood, pp. 129–139. Newcastle upon Tyne, UK: Cambridge Scholars Publishing, 2010.

[52] Cozolino L. The neuroscience of psychotherapy: building and rebuilding the human brain. New York: Norton Professional Books, 2002.

[53] Freud S. (1895) A project for a scientific psychology. Standard Edition, London: Hogarth Press, 1950.

[54] Kandel E. Biology and the future of psychoanalysis: a new intellectual framework for psychiatry. American Journal of Psychiatry, 1999;56:505–524.

[55] Stahl SM. Psychotherapy as an epigenetic 'drug': psychiatric therapeutics target symptoms linked to malfunctioning brain circuits with psychotherapy as well as with drugs. Journal of Clinical Pharmacy and Therapeutics, 2012;37:249–253.

[56] Gregurek R. Psychotherapy in 21st century. Psychiatric Danub, 2012;24 (Supp. 3):13–48.

Chapter 4

Comorbid Mental Disorders in Anxiety Disorders: Genetic Aspects of Bipolar Disorders and of Ethnicity

Yun-Hsuan Chang, Sheng-Yu Lee and Ru-Band Lu

Additional information is available at the end of the chapter

http://dx.doi.org/10.5772/66117

Abstract

Anxiety disorder (AD) is commonly comorbid with other mental illness. It could be a state or trait, controversially. Evidence for an association between alcoholism and anxiety has emerged from clinical studies of patients with alcoholism, and those of patients with anxiety disorders. Alcohol dependence (or abuse) as well as bipolar disorder (BP) is usually comorbid with anxiety disorder and/or depressive disorder, which often coexist and are difficult to distinguish from one another. However, in Han Chinese population, the comorbidity rate either with alcoholism or bipolar disorder was not reported as much high as reported in Caucasians, this finding of comorbidity between anxiety/depressive disorders and alcohol dependence (or abuse) or/and bipolar disorders, possibly at the genetic level, makes the differentiation of their categorical diagnoses in the association study vitally important.

Keywords: comorbidity, ethnics, genetics, mental illness

1. Introduction

Feinstein [1] started to draw attention on patients with one or more than one diagnosable disease and defined "Comorbidity" as "*any distinct additional clinical entity that has coexisted or that may occur during the clinical course of a patient who has the index disease under study.*" In the majority of studies, comorbidity refers to the co-occurrence of at least two different disorders in the same individual. Although the majority of comorbidity research has been at the diagnostic (or syndrome) level (i.e., the presence of co-occurring DSM-IV disorders); another approach is to study the extent to which certain symptoms or symptom patterns tend to co-occur [2–5].

Regardless of types of anxiety disorder (AD), anxiety is one of the most prevalent of all psychiatric disorders in other mental illnesses, such as mood disorders and alcoholism. Different types of mental illnesses commonly comorbid with AD as well as the ethnic role will be reviewed. The comorbidity has gained increasing prominence in psychiatry and psychology in the past few decades [6]. A distinction between two types of comorbidity has been drawn a decade ago by Angold et al. [7].

Reiger et al. [8] reported that approximately 15% of people suffered from AD, according to Diagnostic and Statistical Manual-3rd edition (DSM-III) during their lifetime. Keith et al. (1991) compared the prevalence of AD, including subtypes between community and the institute prevalence rates, and found the lifetime rate of AD was 15% overall, 16% in nursing home residents, 28% in prisoners, and 51% in patients in mental hospitals. Further, Robins et al. [9] studied the prevalence of AD and found that phobia is the most common AD (**Figure 1**).

As the more specific and classified the AD was studied, simple phobia was reported as the most common comorbidity, up to nearly 50%. Approximately 13% of people reported symptoms matching the DSM criteria, and social anxiety disorder was in the second place of highest reported disorder of anxiety. Post-traumatic stress disorder (PTSD), often goes unrecognized, but its prevalence reached 20% in victims of war trauma.

However, the more commonly recognized disorders, such as generalized anxiety disorder (GAD) and panic disorder (PD), have the lower lifetime prevalence rates of approximately 5 and 3.5%, respectively. Another often underdiagnosed disorder, obsession-compulsive disorder (OCD), is found in 2.5% of the population. Interestingly, a recent study found very little change in the prevalence of mental disorders, including specific anxiety disorders, since 1990 [10].

Because of high recognition and high prevalence rate, researchers started to explore whether AD is a genetic-related psychiatric disorder. Therefore, genetic risk factors are being studied; researchers have found genetic predisposition for two broad groups of anxiety disorders: a panic-generalized anxiety-agoraphobia group and a specific phobia group [11]. More clinically important risk factors include comorbid substance abuse and family history. Weissman et al. [12] conducted a 20-year study in the offspring of depressed parents and found a three-fold increase in ADs, including greater substance abuse, younger onset, and more significant physical health concerns.

Although a genetic predisposition for developing an AD is likely [11], environmental stressors clearly play a role in varying degrees. All of the disorders are affected in some way by

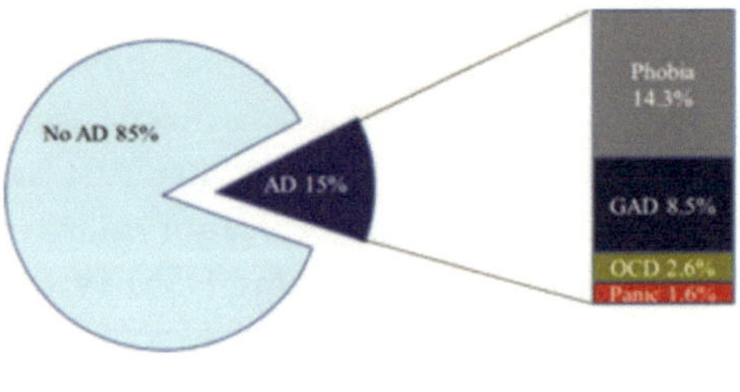

Figure 1. Robins et al. [9] studied in the prevalence of subtypes of AD.

external cues and how they are proceeded and reacted to. Research has also shown that patients suffering from anxiety are generally more sensitive to physiologic changes than non-anxious patients, and panic disorder sufferers are even more sensitive to these than the GAD patients. Objective testing, however, reveals that physiologic changes between anxious and nonanxious patients are comparable. This heightened sensitivity leads to diminished autonomic flexibility, which may be the result of faulty central information processing in anxiety-prone persons [13].

The neuroanatomical foundation of anxiety may be related to the influence of the septohippocampal system in the brain on learning and memory [14]. Patients with AD manifest impaired divided attention [15], verbal learning, verbal recall [16], visual learning and memory [17], episodic memory and executive function [18], and cognitive information processing [19].

2. Anxiety disorder as comorbidity

2.1. Anxiety disorder as comorbidity in bipolar disorders

Prior to year 2000, there were few studies on bipolar disorder II (BPII) and little research into the differences between BPI and BPII patients. More and more studies have been conducted to distinguish between the subdivisions of bipolar disorders (BPI and BPII). Genetically, there was an association with the interaction between *COMT* and *DRD3* gene in BPI [20] as well as the interaction between *DRD2/ANKK1* and the *ALDH2* gene in BPII patients [21]. A statistically significant, main effect for the Met/Met genotype of the *COMT* Val158Met polymorphism and a significant interaction effect for the Met/Met genotype of the *COMT* Val158Met and Ser/Ser genotypes of the *DRD3* Ser9Gly polymorphism could predict BPI, but not BPII compared to normal subjects [20]. Moreover, Lee et al. [22] provided evidence that the *ALDH2* and *5-HT2A* genes interact in BPI but not in BPII. A series of studies have been conducted to prove that BPI and BPII are different from genetic aspect.

Anxiety disorders have been reported as a common comorbidity in BP [23–28], the life time prevalence was reported by National Institute of Mental Health (NIMH) as 51.2% [29, 30], and current anxiety disorders in 31% of the first 500 patients enrolled in the Systematic Treatment Enhancement Program for Bipolar Disorder (STEP-BD) [30]. The National Comorbidity Survey reports that approximately 90% (BPI: 86–92%; BPII: 89%) of bipolar patients comorbid with AD [31–33]. More often, comorbid with panic disorder (PD), post-traumatic stress disorder (PTSD), and obsessive-compulsive disorder (OCD) either in the community or in primary care and psychiatric settings has been reported [30, 32, 34, 35]. There is a growing body of research evidence that bipolar depressives have comparable if not higher rates of comorbid anxiety disorders than unipolar depressives [23, 24, 26–28, 36]. Accordingly, anxiety comorbidity could be a fundamental feature of bipolar disorder [37]. Recent studies reported that among all bipolar disorders, BPII has higher comorbidity with AD than does BPI [23, 25, 26, 28, 38].

Other important aspects to be borne in mind are recognition and prognosis. Compared to noncomorbid BP, BP patients with AD are susceptible to a higher risk of suicidal behavior [39], substance abuse [30, 40, 41], lower psychosocial performance [29] and has a more frequent family history of mental illness [42]. Moreover, the AD comorbidity can complicate the course of illness and pharmacological treatment strategies [43]. Not only the high comorbidity between BP and AD has been noticed, but the types of AD comorbidity in BP could also complicate the recognition of BPI and BPII and predict the prognosis. Patients with BP and AD are susceptible to higher risk of suicidal behaviour [39], substance abuse [30, 40, 41], lower psychosocial performance [29], and a more frequent family history of mental illness than BP patients without AD [42]. Moreover, the AD comorbidity would complicate the course of illness and pharmacological treatment strategies [43].

A number of studies [23–26, 44] have shown bipolar disorder to be highly comorbid with anxiety disorder; a prevalent rate of 51.2% was reported by the National Institute of Mental Health (NIMH) [30, 45]. BP is most frequently associated with panic disorder (PD), post-traumatic stress disorder (PTSD), and obsessive-compulsive disorder (OCD), whether in the community, in primary care, or in psychiatric settings [30, 32, 34, 35]. The National Comorbidity Survey reports that approximately 90% of BP patients (BP-I: 86–92%; BP-II: 89%) have at least one comorbid AD [31–33].

There is a growing body of research evidence that individuals with bipolar depression have comparable if not higher rates of comorbid ADs than do individuals with unipolar depression [23, 24, 26, 27, 36, 44]. Anxiety comorbidity is, then, a fundamental feature of BP [37]. Moreover, established studies [23, 25, 26, 38, 44] show that BP-II is more often comorbid with AD than is BPI. The BP patients with AD comorbidity have been pointed out to have worse prognosis, such as shortened euthymia, delayed remission, and rapid cycling. The AD comorbidity worsens BP patients' episode and their response to treatment, and increases their suicidal behavior. Previous researchers have also pointed out a higher possibility of comorbidity with substance use disorder in BP patient comorbid with AD (BP+*AD*) than in BP without AD (BP^{-AD}) [30, 34, 46–54]. Consequently, the comorbidity with AD receives more attention clinically [55, 56].

2.2. Anxiety disorder as comorbidity in bipolar disorders of ethnics

Chang et al. [57] found that the anxiety disorder comorbidity rate in both BPI (26.7%) and BPII (39.0%) were lower in Han Chinese in Taiwan than in Western populations (more than 50% in BPI and BPII) [26, 30]. There was no significant difference in the gender-based distribution of anxiety disorder in our patients, which agrees with one study [58] but disagrees with others [34, 59, 60] that report a higher prevalence in women than in men. One reason may be that lower prevalence of a disease shows a greater statistical meaning for fixed heritability and a fixed number of trait loci [61]. Therefore, the lower anxiety disorder comorbidity rate not only is more easy for researchers to look at the genetic factor of AD high comorbid with BP but also may decrease psychosocial, cultural, and other confounding factors. Those factors and their interaction might increase the prevalence rate of mental illness, especially anxiety disorder. In addition, a higher comorbidity rate with BPII than with BPI was found which agreed with most of large-sample epidemiological studies [23, 25, 26, 28, 38]. Patients with BP and co-occurring anxiety symptoms or anxiety disorders are susceptible to higher rates of depressive episodes [29], which may explain the higher comorbidity rate in patients with BPII than with BPI.

For further investigation of the AD subtypes with BP, Chang et al. [57] found that the highest AD comorbidity in both subtypes of BP patients was GAD instead of PD or OCD, the major AD comorbidities in the western BP populations. Wittchen et al. [62] reported that the highest rate of comorbidity of GAD was associated with major depressive disorder (62.4%) and the lowest with BP (10.5%). One of the possible reasons may be due to ethnic differences or the genetic heterogeneity of anxiety disorder and depressive disorder like BPII. A higher occurrence of GAD with BPII was found in Chang et al. (2012)'s study, which is similar to previous studies showing higher occurrence of GAD with MDD [63, 64]. This gender difference in BPII could be derived from the high GAD-associated depression. However, the causal relationship should be further investigated. The other reason could be that the same diagnostic criteria of BPI and BPII except the duration might not be appropriate, the redefinition of the diagnostic criteria after more different ethnic BP subtype studies are suggested.

Chang et al. [57] have reported a relatively low rate of anxiety-disorder comorbidity in both BP subtypes in Han Chinese population in Taiwan, implying an ethnic possibility. Because of this low anxiety-disorder comorbidity in BP population in Taiwan, it was easier to identify BP patients with/without comorbidities to study the influence of anxiety-disorder comorbidity on neuropsychiatric performance. Although causal relationship between the comorbid anxiety disorders in BPI and BPII is not yet clear, additional studies are required, and the ethnic differences are suggested to be taken into account.

BPII patients with anxiety-disorder comorbidity also showed more substance abuse and dependence, suicide attempts, and personality disorders than did BPI patients [23, 25, 26, 28, 30, 57, 65]. After excluding the comorbidity of anxiety disorder, BPI and BPII patients have similar suicide rates, suggesting that AD comorbidity increases the risk of suicide in BPII [66]. The risk of suicide in Han Chinese BP patients in Taiwan may be lower than western BP populations, but it needs further study to confirm.

Anxiety disorders occur most frequently during depressive episodes in patients with BP [67], except for those in the depressive state, anxiety disorder would often present during subsyndromal states [68]. Boylan et al. [49] reported that about 32% of their BP patients had more than two comorbid anxiety disorders. The lower incidence in BPI patients (21.9%) and a similar incidence in BPII patients (33.8%) were found. This finding implies that BP patients with multiple anxiety-disorder comorbidities may have a more severe psychopathology and a worse prognosis, even with what is currently considered as appropriate treatment [69]. In addition, anxiety disorder may be a predisposing factor for BP. Long-term follow-up studies are needed to confirm whether some anxiety-disorder comorbidities are in remission during the inter-episode stage or increase in intensity as symptoms of depression worsen.

2.3 Genetic aspect in comorbidity with anxiety disorder in bipolar disorders

Relative genetic factor between BP and AD was proposed [70], but no definite susceptibility gene for BP has been identified. One possible explanation could be neither subtypes of BP nor comorbidities were differentiated in most studies [71–73]. The contribution of genetic factors to the etiology of BP has also been reported from studies on family, twin, and adoption [72, 74]. From the twin studies, the inheritance for bipolar disorder overall is around 85% [75, 76].

The distinction between BPI and BPII may be associated with different genetic categories [21, 77–79]. Knowing the association between AD comorbidity and BP at the genetic level may improve our understanding more in this mental illness.

Dopaminergic dysfunction has been implicated in the pathogenesis of bipolar disorder, especially the dopamine D2 receptor gene (*DRD2*) on chromosome 11q22.3, and dopamine D3 receptor gene (*DRD3*) on chromosome 3q13.3, expressed exclusively in the limbic regions of the brain responsible for controlling emotions and behavior as well as cognition [80, 81]. The DRD2 Taq-IA (rs1800497) restriction in fragment length of polymorphism is linked to the density of dopamine D2 receptors [82]. A variety of studies have analyzed dopamine receptor genes for their associations with BP or AD, including polymorphisms in DRD2 [83, 84]. The *DRD3* gene is encoded as a target site for antipsychotic agents, which are efficient in the treatment of this disorder. The most frequently studied allele variation of the *DRD3* gene is the *DRD3* Ser9Gly polymorphism (rs6280) [85], causing a serine- (Ser) to-glycine (Gly) substitution and a significantly increased dopamine binding affinity [86]. The Gly9 allele is associated with significantly greater odds for treatment response to antipsychotics [87, 88]. Possible association between the *DRD2* gene, *DRD3* gene, and bipolar disorder has been reported in a family-based study [89] but not in others [83, 84, 90, 91]. Therefore, the *DRD2 and DRD3* genes are of particular interest in the study of susceptibility to bipolar disorder because both effective episodes and neurocognitive impairments are important aspects of BP [92] and AD.

In the midbrain-hindbrain regions, another important role in the development of dopaminergic neuron, brain-derived neurotrophic factor (BDNF), in which the *BDNF* gene was selectively deleted, the number of tyrosine hydroxylase-expressing dopaminergic neurons was found reduced [93]. During the developing years and in adulthood, *BDNF* gene encoded on human chromosome 11p13 was shown to regulate the expression of DRD3 in the nucleus accumbens [94]. The involvement of BDNF in the pathogenesis of mood disorders and the mechanism of mood stabilizing medication has been suggested [95]. Genetic studies explored the potential association between *BDNF* gene variants and bipolar disorders yet yield conflicting results. The most investigated *BDNF* gene is Val66Met (rs6265) with functional consequences from valine (Val) to methionine (Met) at codon 66 [95–98]. The Met allele has been pointed out to have association with impairments in intracellular trafficking and activity-dependent secretion of BDNF in neurons [95–98]. A significant association between Val66Met and bipolar disorder has also been reported in several studies conducted in North American and European populations [100–103] with overtransmission of the Val allele while other studies in the Asian populations were not [99–101]. About 80% with Val allele was reported in the European population while only 50% was reported in Asian [102], the ethnic difference could be the possibility.

2.4. The DRD2 gene associated with ALDH2 in bipolar II disorder with anxiety disorder

There is an increased risk of mental disorder among relatives of anxiety neurotics from family studies [103]. Therefore, if the *DRD2* locus is linked to a predisposition to conduct AD and BP, its relation with anxiety disorder and BP is worth further examination. Knutson et al. [104] proposed molecular explanation that a low serotonin turnover rate and aggressive behavior are mediated by negative emotions such as insecurity and anxiety. Some studies also suggest that conduct disorder might lead to alcoholism because of the tendency for a person to be impulsive [105] and to exhibit behavior disinhibition [106]. Besides the association of *DRD2* gene with anxiety and mood regulation, we also set forth to determine the possible relationship with *DRD2* in AD and BP to examine any association between *DRD2* and other possible genetics.

Brain-imaging study [82] shows that healthy controls with an *A1* allele of the *DRD2 TaqIA* gene have fewer *DRD2* receptors than those without the *A1* allele. Individuals with at least one *A1* allele appear to have up to 40% fewer striatal DRD2 receptors than those carrying the A2/A2 allele [107]. The *DRD2 TaqIA A1* allele was associated with fewer DRD2s in the striatum and hence, a lower dopaminergic function.

Enzymes that function in the metabolic breakdown of acetaldehyde are considered as the *ALDH2* gene; the enzyme functioning in the metabolic process of acetaldehyde is majorly influencing drinking behavior and the development of alcoholism. The ALDH2 shows two variant alleles: *ALDH2*1* and *ALDH2*2*. The *ALDH2*1/*1*-encoded enzyme is active in the metabolism of acetaldehyde, whereas the enzymes encoded by the *ALDH2*1/*2* and *ALDH2*2/*2* are partially and totally inactive, respectively. It is believed that the *ALDH2*2* allele, with reduced enzyme activity, provides protection against the risk of developing alcoholism [108]. In previous reports, nearly half of the East Asian population has the *ALDH2*2* allele variant [109], including Han Chinese in Taiwan [108, 110], but this allele is rarely found in other ethnic populations.

Wang et al. [111] revealed the relationship among *DRD2* gene and *ALDH2* gene between BP with and without AD and found, we examined whether the DRD2 and ALDH2 genes were associated with comorbid BP-II and AD. The study results revealed the significant association of DRD2 Taq-IA A1/A2 in the BP-II with AD and the significant interaction between the ALDH2 and DRD2 genes in BP-II without AD, respectively. Our findings provide genetic evidence to support the hypothesis that BPII with or without AD might be two distinct mental illnesses [112, 113]. Such interaction also implies the complex role of dopamine system in the pathogenesis of BPII.

2.5. The DRD3 gene and BDNF gene associated with bipolar subtypes with/out anxiety disorder

Chang et al. [114] investigated the association between *DRD3* gene and *BDNF* gene in bipolar subtypes comorbid with/without AD. They found a significant main effect of the Ser/Gly of the *DRD3* Ser9Gly polymorphism in $BPII^{+AD}$, and interaction with the *BDNF* Val66Met Met/Val genotype and indicated a possible mediator by the *BDNF* Val66Met Val/Val genotype in the development to the AD comorbidity in the BPII. The involvement of the *DRD3* Ser9Gly Ser/Gly genotype disagrees with the previous studies of the risk to develop anxiety disorders, such as obsessive compulsive personality (OCD) [115, 116]. This difference may reflect the report that in Han Chinese, the most AD subtype comorbidity in BP subtypes is generalized anxiety disorder [57] instead of OCD or panic disorder mostly reported in the western population [6]. Further investigation between ethnics should be considered.

Takahashi et al. [117] has reported the effect of the *DRD3* and *BDNF* variation on brain morphology in midline and medial temporal lobe structures in healthy controls. Gourion et al. [118] earlier has reported that this interaction was associated with earlier emergence of psychosis in schizophrenia patients. The involvement of abnormal dopamine regulation in bipolar disorders have been reported in gene-gene interaction study; however, this interaction is associated with treatment response of dopamine receptor antagonists for bipolar disorders or previous psychotic symptoms may require further studies with additional characterization and phenotyping.

There have been some reports about that the Met heterozygotes compared to the Val homozygotes of the *BDNF* Val66Met polymorphism have impairment in intracellular trafficking and activity-dependent secretion of BDNF in neurons [95–98]. In addition, the Ser heterozygotes compared to the Gly homozygotes of the *DRD3* Sere9Gly polymorphism were reported to decrease the dopamine binding affinity [86]. Odds for BPII^{+AD} were higher for those BDNF Met66Val Met/Met and Met/Val with the interaction of the *DRD3* Ser9Gly Ser/Gly genotype compared to those with Gly/Gly genotypes.

For the results in the BPI, a main effect of *BDNF* Met/Val genotype was found to be associated with BPI^{-AD}, and an interaction was found between *BDNF* Met/Val with *DRD3* Ser9Gly Ser/Ser genotype in BPI^{-AD}. From the results in both subtypes, it implies the different involvement of *DRD3* and *BDNF* genetic variants between two subtypes of the BP, and the variation of neuropsychological performance [119]. No case with the *DRD3* Ser9Gly Gly/Gly genotype was reported in the BPI^{-AD}, and impact of the AD comorbidity on the BPI impaired their executive function and attention, while psychomotor speed, working memory, and visual immediate memory was impaired in the BPII [119], implying the different roles of the *DRD3* Ser9Gly polymorphisms.

Chang et al. [114] have provided initial evidence of the involvement of dopaminergic pathway with *DRD3 Ser9Gly* gene in the pathogenesis of bipolar disorder. For the results in the BPI, a main effect of *BDNF* Met/Val genotype was found to be associated with BPI^{-AD}, and an interaction was found between *BDNF* Met/Val with *DRD3* Ser9Gly Ser/Ser genotype in BPI^{-AD}. From the results in both subtypes, it implies the different involvement of *DRD3* and *BDNF* genetic variants between two subtypes of the BP, and the variation of neuropsychological performance [119]. No case with the *DRD3* Ser9Gly Gly/Gly genotype was reported in the BPI^{-AD}, and impact of the AD comorbidity on the BPI impaired their executive function and attention, while psychomotor speed, working memory, and visual immediate memory was impaired in BPII [119], implying the different roles of the *DRD3* Ser9Gly polymorphisms.

The results in Chang et al.'s study [114] not only replicated Lohoff et al. [120] finding that the positive association was between the Val allele and the BPI patients, but also related to the AD comorbidity. However, a disagreement was noticed between this study and previous findings of association between the Met allele with anxiety disorders [121, 122], indicating the possible ethnic variation. Moreover, the major subtype of AD comorbidity in the Han Chinese BPI and BPII was general anxiety disorder [57] while PD or OCD was the higher comorbidity with the BP in the western population [30, 32, 34, 35]. The genotype distribution of the *BDNF* Val66Met polymorphism in current study is consistent with other Asian populations, but different from European populations [102].

3. Anxiety disorder as comorbidity in alcoholism of ethnics

3.1. The relationship between DRD2 gene and ALDH2 gene in anxiety-depression alcoholism

Several studies in alcohol detoxification have been reported that the *DRD2* gene may be associated with high scores of anxiety and depression in alcohol dependence. Patients with alcohol dependence comorbid with anxiety and/or depressive disorders represented the greatest risk of relapse [123–125]. In a double-blind treatment study using bromocriptine, ALC individuals with *DRD2 A1* allele showed the greatest improvement in craving and anxiety [126]. These observations are in agreement of an association between *DRD2* gene and anxiety-depressive alcohol dependence (or abuse). To elucidate the association between the *DRD2* gene and alcohol dependence in Han Chinese population with attempts to overcome the possible confounding effects and to reduce false-positive or -negative results, Huang et al. [127] compared individuals with solely anxiety-depression (ANX/DEP), individuals with both alcohol dependence and anxiety-depression (ANX/DEP ALC), and individuals with pure alcohol dependence and normal controls. Strong linkage disequilibrium between the *Taq*I A and B polymorphisms of the *DRD2* gene was reported. Huang et al. [127] found that the frequency of the *A1/B1* haplotype was significantly higher in the ANX/DEP ALC group than that of controls. There was no association between the *DRD2* haplotype and pure alcohol dependence or ANX/DEP when compared to controls.

Since ALDH2 is a crucial enzyme for ethanol catabolism which might also play an important role in dopamine catabolism and risk for alcoholism, the involvement of the *ALDH2* gene with the association of the *DRD2* gene and ANX/DEP ALC was further investigated. It was shown that the *DRD2* gene is associated with ANX/DEP ALC only after controlling for the *ALDH2*1/*1* genotype, supporting the contention that the *DRD2* gene may interact with the *ALDH2* genes in ANX/DEP ALC.

3.2. Interaction between personality traits and genes in anxiety-depressive alcoholism

Cloninger has hypothesized of lower novelty seeking and higher harm avoidance in type I alcoholism compared with healthy volunteers. Later, Huang et al. [127] and Huang et al. [128] have proved that anxiety-depressive alcohol dependence (ANX/DEPALC) could be a genetically well-defined subtype of alcoholism linking to *DRD2*. Similar clinical characteristics with Type I alcoholism has been found in the ANX/DEPALC with late-onset and more anxious/depressed traits, and their suffering from anxiety/depression related to heavy drinking. The higher NS and HA scores were found in the ANX/DEPALC than in the pure ALC. This result might indicate that ANX/DEP ALC belongs to a subtype of alcoholism [127]. Furthermore, an association was found between ANX/DEP ALC and NS, but only in subjects with *DRD2 TaqIA* A1+ allele (including A1/A1 or A1/A2 genotype). In addition, the difference in NS between ANX/DEP ALC and Pure ALC existed in subjects with S/S genotype of *5-HTTLPR*. The potential genes, *DRD2 TaqIA* A1+ allele and 5-HTTLPR may involve in the development of ANX/DEP ALC with novelty seeking personality trait [127].

Further analysis with stratification of the *DRD2 TaqIA* A1/A1 or A1/A2 genotype subjects, the difference in NS scores was only found in subjects with *5-HTTLPR* S/S genotype. The ANX/DEPALC was associated with HA only in subjects with *5-HTTLPR* S/L and L/L genotypes, suggesting that the personality traits of type I alcoholism in Cloninger's model might need modification. The *5-HTTLPR* polymorphism involved in both NS and HA implied that personality traits related to multiple genes could be possible in developing mental disorder. Therefore, multiple genes would be suggested to be considered in the further study .

4. Summary

Wang et al. [111] provided preliminary evidence that comorbid BP-II^{+AD} might be one of the subtypes of BPII. The *DRD2* gene could be an important candidate gene for the comorbidity of AD in other mental illness and *ALDH2* gene might moderate the impact of *DRD2* gene on BPII with or without AD. Moreover, Chang et al. [114] revealed the exact effect of this interaction on DRD3 binding affinity and neuron secretion of BDNF is not clear yet may be associated with the pathogenesis of anxiety disorders. The Gly/Gly genotypes of the *DRD3* Ser9Gly have been associated with unipolar depression [129]; this might explain the main effect was only found in the BPII^{+AD} but not in the BPI^{+AD}. A possibility could be ethnic differences or the genetic heterogeneity of AD and depressive disorder like BPII; because in the Han Chinese, GAD has been reported as the major AD comorbidity in the BPII [57] while it was reported to be associated with major depression in other ethnics [64, 130]. In addition, ADs occur most frequently during depressive and depressive manic episodes in BP patients [67] as well as during subsyndromal depressive states [68]. Since long-term follow-ups have shown that patients with BPII have a more chronic course, more mood episodes, more major and minor depressive episodes which last longer than those of patients with BPI [131–133], whether the interaction between the *BDNF* and *DRD3* genes is related to the clinical characteristics of BPII^{+AD}, for example, depression-proneness, may still require further study.

Previous studies provided initial evidence of the involvement of dopaminergic pathway as well as serotonin system in the pathogenesis of bipolar disorder [110, 113, 134, 135]. However, whether the interaction of these genes leads to dysfunctional dopaminergic signaling or serotonin regulation and to what extent these genes would affect the etiology of bipolar disorder with AD comorbidity still require further clarification. Moreover, AD playing the role of comorbidity in BP and alcoholism showed relatively lower prevalence rate in Han Chinese while compared to previous studies conducted in Western population. This finding implied an ethnic possibility which has been supported in some studies with genetic investigation. Moreover, in both BP and ALC with AD comorbidity have been reported to have relationship with dopaminergic genes as well as serotonin-related genes, the *ALDH2* gene plays the most important role in both disorders with AD comorbidity. The frequency of *ALDH2* gene has been found differently in Han Chinese compared to Caucasians, further causal-relationship investigation of this gene in AD comorbidity is needed to confirm.

http://dx.doi.org/10.5772/66117

Abbreviations

AD: anxiety disorder

ALDH2: aldehyde dehydrogenase 2

ANX/DEP ALC: anxiety/depression alcoholism

BDNF: brain-derived neurotrophic factor

BP: bipolar disorder

DOPAC: 3,4-dihydroxyphenylacetic acid

DOPAL: 3,4-dihydroxyphenylacetaldehyde

DRD2: dopamine D2 receptor

DRD3: dopamine D3 receptor

GAD: generalized anxiety disorder

PTSD: post-traumatic stress disorder

PD: panic disorder

OCD: obsession compulsive disorder

MAOA: monoamine oxidase A

Author details

Yun-Hsuan Chang[1,2,3], Sheng-Yu Lee[3,4] and Ru-Band Lu[3,5]*

*Address all correspondence to: rblu@mail.ncku.edu.tw

1 Department of Psychology, College of Medical and Health Sciences, Asia University, Taichung, Taiwan, R.O.C

2 Department of Medical Research, China Medical University Hospital, China Medical University, Taichung, Taiwan, R.O.C

3 Department of Psychiatry, National Cheng Kung University, Tainan, Taiwan, R.O.C.

4 Department of Psychiatry, Kaohsiung Veteran's General Hospital, Kaohsiung, Taiwan, R.O.C

5 Institute of Behavioural Sciences, College of Medicine, National Cheng Kung University, Tainan, Taiwan, R.O.C.

References

[1] Feinstein AR. The pre-therapeutic classification of co-morbidity in chronic disease. Journal of Chronic Diseases. 1970;23(7):455–468.

[2] Blashfield RK, Sprock J, Fuller AK. Suggested guidelines for including or excluding categories in the DSM-IV. Comprehensive psychiatry. 1990 Jan-Feb;31(1):15–19. PubMed PMID: 2297982. Epub 1990/01/01. eng.

[3] Boyd JH, Burke JD, Jr., Gruenberg E, Holzer CE, 3rd, Rae DS, George LK, et al. Exclusion criteria of DSM-III. A study of co-occurrence of hierarchy-free syndromes. Archives of General Psychiatry. 1984 Oct;41(10):983–989. PubMed PMID: 6477056. Epub 1984/10/01. eng.

[4] Di Nardo PA, Barlow DH. Syndrome and symptom comorbidity in the anxiety disorders. In: Maser JDC, C. R., editor. Comorbidity in anxiety and mood disorders. Washington, DC: American Psychiatric Press; 1990. pp. 205–230.

[5] Moras K, Barlow D. Definitions of secondary depression: effects on comorbidity and outcome in anxiety disorders. Psychopharmacol Bull. 1992;28(1):27–33. PubMed PMID: 1609039. Epub 1992/01/01. eng.

[6] Wittchen HU. Critical issues in the evaluation of comorbidity of psychiatric disorders. British Journal of Psychiatry Supplement. 1996;30:9–16.

[7] Angold A, Costello EJ, Erkanli A. Comorbidity. Journal of Child Psychology Psychiatry. 1999 Jan;40(1):57–87. PubMed PMID: 10102726. Epub 1999/04/02. eng.

[8] Reiger DA, Burker JD, Burke KC. Comorbidity of affective and anxiety disorders in the NIMH Epidemiological Catchment Area Program. In: Maser JD, Cloniger CR, editors. Comorbidity of mood and anxiety disorders. Washington, DC: American Psychiatric Press; 1990. pp. 113–122.

[9] Robins LN, Locke BZ, Regier DA. An overview of psychiatric disorders in America. In: Robbins LN, Regier DA, editors. Psychiatric disorders in America: the epidemiologic catchment area study. New York: Free Press; 1991. pp. 328–366.

[10] Kessler RC, Demler O, Frank RG, Olfson M, Pincus HA, Walters EE, et al. Prevalence and treatment of mental disorders, 1990 to 2003. New England Journal of Medicine. 2005 Jun 16;352(24):2515–2523. PubMed PMID: 15958807. Pubmed Central PMCID: 2847367. Epub 2005/06/17. eng.

[11] Hettema JM, Prescott CA, Myers JM, Neale MC, Kendler KS. The structure of genetic and environmental risk factors for anxiety disorders in men and women. Archives of General Psychiatry. 2005 Feb;62(2):182–189. PubMed PMID: 15699295. Epub 2005/02/09. eng.

[12] Weissman MM, Wickramaratne P, Nomura Y, Warner V, Pilowsky D, Verdeli H. Offspring of depressed parents: 20 years later. The American Journal of Psychiatry. 2006 Jun;163(6):1001–1008. PubMed PMID: 16741200. Epub 2006/06/03. eng.

[13] Hoehn-Saric R, McLeod DR, Funderburk F, Kowalski P. Somatic symptoms and physiologic responses in generalized anxiety disorder and panic disorder: an ambulatory monitor study. Archives of General Psychiatry. 2004 Sep;61(9):913–921. PubMed PMID: 15351770. Epub 2004/09/08. eng.

[14] Gray JA. The neuropsychology of anxiety: An enquiry into the functions of the septo-hippocampal system. Oxford: Oxford University Press; 1982.

[15] Lautenbacher S, Spernal J, Krieg JC. Divided and selective attention in panic disorder. A comparative study of patients with panic disorder, major depression and healthy controls. European Archives of Psychiatry and Clinical Neuroscience. 2002 Oct;252(5):210–213. PubMed PMID: 12451461. Epub 2002/11/27. eng.

[16] Asmundson GJ, Stein MB, Larsen DK, Walker JR. Neurocognitive function in panic disorder and social phobia patients. Anxiety. 1994;1(5):201–207. PubMed PMID: 9160575. Epub 1994/01/01. eng.

[17] Lucas JA, Telch MJ, Bigler ED. Memory functioning in panic disorder: A neuropsychological perspective. Journal of Anxiety Disorders. 1991;5(1):1–20.

[18] Airaksinen E, Larsson M, Forsell Y. Neuropsychological functions in anxiety disorders in population-based samples: evidence of episodic memory dysfunction. Journal of Psychiatric Research. 2005 Mar;39(2):207–214. PubMed PMID: 15589570. Epub 2004/12/14. eng.

[19] Sachs G, Anderer P, Margreiter N, Semlitsch H, Saletu B, Katschnig H. P300 event-related potentials and cognitive function in social phobia. Psychiatry Research. 2004 Sep 15;131(3):249–261. PubMed PMID: 15465294. Epub 2004/10/07. eng.

[20] Lee SY, Chen SL, Chen SH, Huang SY, Tzeng NS, Chang YH, et al. The COMT and DRD3 genes interacted in bipolar I but not bipolar II disorder. World Journal of Biological Psychiatry. 2010 Aug 10 ; 12(5):385-91. PubMed PMID: 20698735. Epub 2010/08/12. Eng.

[21] Lee SY, Chen SL, Chang YH, Chen SH, Chu CH, Huang SY, et al. The ALDH2 and DRD2/ANKK1 genes interacted in bipolar II but not bipolar I disorder. Pharmacogenet Genomics. 2010 Aug;20(8):500–506. PubMed PMID: 20577142. Epub 2010/06/26. eng.

[22] Lee SY, Chen SL, Chang YH, Chu CH, Huang SY, Tzeng NS, et al. The ALDH2 and 5-HT2A genes interacted in bipolar-I but not bipolar-II disorder. Progress in Neuro-Psychopharmacology & Biological Psychiatry. 2012 Aug 7;38(2):247–251. PubMed PMID: 22564712. Epub 2012/05/09. eng.

[23] Angst J. Panic disorder: History and epidemiology. European Psychiatry: The Journal of the Association of European Psychiatrists. 1998;13 (Suppl 2):51s–55s. PubMed PMID: 19698673.

[24] Chen YW, Dilsaver SC. Comorbidity of panic disorder in bipolar illness: evidence from the Epidemiologic Catchment Area Survey. American Journal of Psychiatry. 1995;152:280–282.

[25] Endicott J, Nee J, Andreasen N, Clayton P, Keller M, Coryell W. Bipolar II Combined or keep separate? Journal of Affective Disorder. 1985;8:17–28.

[26] Rihmer Z, Szadoczky E, Furedi J, Kiss K, Papp Z. Anxiety disorders comorbidity in bipolar I, bipolar II and unipolar major depression: results from a population-based study in Hungary. Journal of Affective Disorders. 2001 Dec;67(1–3):175–179. PubMed PMID: 11869765. Epub 2002/03/01. eng.

[27] Manning JS, Haykal, RF, Connor, PD. On the nature of depressive and anxious states in family practice setting: the high prevalence of bipolar II and related disorders in a cohort followed longitudinally. Comprehensive Psychiatry. 1997;38:102-108.

[28] Perugi G, Akiskal, HS, Ramaciotti, S, Nassini, S, Toni, C, Milanfranchi, A, Musetti, L. Depressive comorbidity of panic, social phobia and obsessive-compulsive disorders re-examined: is there a bipolar II connection? Journal of Psychiatric Research. 1999;33:53–61.

[29] Bauer M, Altshuler L, Evans D, Beresford T, Williford W, Hauger R, et al. Prevalence and distinct correlates of anxiety, l substance, and combined comorbidity in a multi-site public sector sample with bipolar disorder. Journal of Affective Disorder. 2005;85:301–315.

[30] Simon NM, Otto MW, Wisniewski SR, Fossey M, Sagduyu K, Frank E, et al. Anxiety disorder comorbidity in bipolar disorder patients: data from the first 500 participants in the Systematic Treatment Enhancement Program for Bipolar Disorder (STEP-BD). The American Journal of Psychiatry. 2004 Dec;161(12):2222–2229. PubMed PMID: 15569893. Epub 2004/12/01. eng.

[31] Kessler RC, DuPont RL, Berglund P, Wittchen HU. Impairment in pure and comorbid generalized anxiety disorder and major depression at 12 months in two national surveys. The American Journal of Psychiatry. 1999 Dec;156(12):1915–1923. PubMed PMID: 10588405. Epub 1999/12/10. eng.

[32] Kessler RC, Rubinow DR, Holmes C, Abelson JM, Zhao S. The epidemiology of DSM-III-R bipolar I disorder in a general population survey. Psychological Medicine. 1997 Sep;27(5):1079–1089. PubMed PMID: 9300513. Epub 1997/09/23. eng.

[33] Merikangas KR, Akiskal HS, Angst J, Greenberg PE, Hirschfeld RM, Petukhova M, et al. Lifetime and 12-month prevalence of bipolar spectrum disorder in the National Comorbidity Survey replication. Archives of General Psychiatry. 2007 May;64(5):543–552. PubMed PMID: 17485606. Pubmed Central PMCID: PMC1931566. Epub 2007/05/09. eng.

[34] Freeman MP, Freeman SA, McElroy SL. The comorbidity of bipolar and anxiety disorders: prevalence, psychobiology, and treatment issues. Journal of Affective Disorders. 2002 Feb;68(1):1–23. PubMed PMID: 11869778.

[35] Yerevanian BI, Koek RJ, Ramdev S. Anxiety disorders comorbidity in mood disorder subgroups: data from a mood disorders clinic. Journal of Affective Disorders. 2001 Dec;67(1-3):167–173. PubMed PMID: 11869764. Epub 2002/03/01. eng.

[36] Chen YW, Dilsaver SC. Comorbidity for obsessive-compulsive disorder in bipolar and unipolar disorders. Psychiatry Research. 1995 Nov 29;59(1–2):57–64. PubMed PMID: 8771221. Epub 1995/11/29. eng.

[37] McIntyre R, Katzman M. The role of atypical antipsychotics in bipolar depression and anxiety disorders. Bipolar Disorders. 2003;5 (Suppl 2):20–35. PubMed PMID: 14700010. Epub 2004/01/01. eng.

[38] Zutshi A, Reddy YC, Thennarasu K, Chandrashekhar CR. Comorbidity of anxiety disorders in patients with remitted bipolar disorder. European Archives of Psychiatry and Clinical Neuroscience. 2006 Oct;256(7):428–436. PubMed PMID: 16783496. Epub 2006/06/20. eng.

[39] Vieta E, Colom F, Martinez-Aran A, Benabarre A, Reinares M, Gasto C. Bipolar II disorder and comorbidity. Comprehensive Psychiatry. 2000;41:339–343.

[40] Dunner DL. Correlates of suicidal behavior and lithium treatment in bipolar disorder. Journal of Clinical Psychiatry. 2004;65(Suppl 10):5–10.

[41] Krishnan KR. Psychiatric and medical comorbidities of bipolar disorder. Psychosomatic Medicine. 2005 Jan-Feb;67(1):1–8. PubMed PMID: 15673617. Epub 2005/01/28. eng.

[42] Wozniak J, Biederman J, Monuteaux M, Richards J, Faraone SV. Parsing the comorbidity between bipolar disorder and anxiety disorders: A familial risk analysis. Journal of Child and Adolescent Psychopharmacology. 2002;12:101–111.

[43] Lee JH, Dunner DL. The effect of anxiety disorder comorbidity on treatment resistant bipolar disorders. Depression and Anxiety. 2008;25(2):91–97. PubMed PMID: 17311265. Epub 2007/02/22. eng.

[44] Perugi G, Akiskal HS, Ramacciotti S, Nassini S, Toni C, Milanfranchi A, et al. Depressive comorbidity of panic, social phobic, and obsessive-compulsive disorders re-examined: is there a bipolar II connection? Journal of Psychiatric Research. 1999 Jan–Feb;33(1):53–61. PubMed PMID: 10094240. Epub 1999/03/27. eng.

[45] Bauer MS, Altshuler L, Evans DR, Beresford T, Williford WO, Hauger R. Prevalence and distinct correlates of anxiety, substance, and combined comorbidity in a multi-site public sector sample with bipolar disorder. Journal of Affective Disorders. 2005 Apr;85(3):301–315. PubMed PMID: 15780700. Epub 2005/03/23. eng.

[46] Post RM. The impact of bipolar depression. Journal of Clinical Psychiatry. 2005;66 (Suppl 5):5–10. PubMed PMID: 16038596. Epub 2005/07/26. eng.

[47] McIntyre RS, Soczynska JK, Bottas A, Bordbar K, Konarski JZ, Kennedy SH. Anxiety disorders and bipolar disorder: a review. Bipolar Disorders. 2006 Dec;8(6):665–676. PubMed PMID: 17156153. Epub 2006/12/13. eng.

[48] McElroy SL, Altshuler LL, Suppes T, Keck PE, Jr., Frye MA, Denicoff KD, et al. Axis I psychiatric comorbidity and its relationship to historical illness variables in 288 patients with bipolar disorder. American Journal of Psychiatry. 2001 Mar;158(3):420–426. PubMed PMID: 11229983. Epub 2001/03/07. eng.

[49] Boylan KR, Bieling PJ, Marriott M, Begin H, Young LT, MacQueen GM. Impact of comorbid anxiety disorders on outcome in a cohort of patients with bipolar disorder. Journal of Clinical Psychiatry. 2004 Aug;65(8):1106–1113. PubMed PMID: 15323597. Epub 2004/08/25. eng.

[50] Bauer MS, Altshuler L, Evans DR, Beresford T, Williford WO, Hauger R. Prevalence and distinct correlates of anxiety, substance, and combined comorbidity in a multi-site public sector sample with bipolar disorder. Journal of Affective Disorders. 2005 Apr;85(3):301–315. PubMed PMID: 15780700. Epub 2005/03/23. eng.

[51] Feske U, Frank E, Mallinger AG, Houck PR, Fagiolini A, Shear MK, et al. Anxiety as a correlate of response to the acute treatment of bipolar I disorder. American Journal of Psychiatry. 2000 Jun;157(6):956–962. PubMed PMID: 10831476. Epub 2000/06/01. eng.

[52] Simon GE. Social and economic burden of mood disorders. Biological Psychiatry. 2003 Aug 1;54(3):208–215. PubMed PMID: 12893097. Epub 2003/08/02. eng.

[53] Simon NM, Zalta AK, Otto MW, Ostacher MJ, Fischmann D, Chow CW, et al. The association of comorbid anxiety disorders with suicide attempts and suicidal ideation in outpatients with bipolar disorder. Journal of Psychiatric Research. 2007 Apr-Jun;41(3–4):255–64. PubMed PMID: 17052730. Epub 2006/10/21. eng.

[54] Vojta C, Kinosian B, Glick H, Altshuler L, Bauer MS. Self-reported quality of life across mood states in bipolar disorder. Comprehensive Psychiatry. 2001 May–Jun;42(3):190–5. PubMed PMID: 11349236. Epub 2001/05/12. eng.

[55] Wittchen HU. Generalized anxiety disorder: prevalence, burden, and cost to society. Depression and Anxiety. 2002;16(4):162–71. PubMed PMID: 12497648. Epub 2002/12/24. eng.

[56] Kessler RC. The epidemiology of pure and comorbid generalized anxiety disorder: a review and evaluation of recent research. Acta Psychiatric Scandinavica Supplement. 2000; 406:7–13. PubMed PMID: 11131470. Epub 2000/12/29. eng.

[57] Chang YH, Chen SL, Chen SH, Chu CH, Lee SY, Yang HF, et al. Low anxiety disorder comorbidity rate in bipolar disorders in Han Chinese in Taiwan. Progress in Neuro-Psychopharmacology & Biological Psychiatry. 2012 Jan 10;36(1):194–197. PubMed PMID: 21996277. Epub 2011/10/15. eng.

[58] Kawa I, Carter J, Joyce P, Doughty C, Frampton C, Wells J, et al. Gender differences in bipolar disorder: Age of onset, course, comorbidity, and symptom presentation. Bipolar Disorders. 2005;7:119–125.

[59] Kessler RC, McGonagle KA, Zhao S, Nelson CB, Hughes M, Eshleman S, et al. Lifetime and 12-month prevalence of DSM-III-R psychiatric disorders in the United States. Results from the National Comorbidity Survey. Archives of General Psychiatry. 1994 Jan;51(1):8–19. PubMed PMID: 8279933. Epub 1994/01/01. eng.

[60] Strakowski SM, Tohen M, Stoll A, Faedda G, Goodwin D. Comorbidity in mania at first hospitalization. American Journal of Psychiatry. 1992;149:554–556.

[61] Rice J, Saccone N, Rasmussen E. Definition of the phenotype. In: Rao D, Province M, editors. Genetic discussion of complex traits. New York: Academic Press; 2001. pp. 70–74.

[62] Wittchen HU, Zhao S, Kessler RC, Eaton WW. DSM-III-R generalized anxiety disorder in the National Comorbidity Survey. Archives of General Psychiatry. 1994 May;51(5):355–364. PubMed PMID: 8179459. Epub 1994/05/01. eng.

[63] Moffitt TE, Harrington H, Caspi A, Kim-Cohen J, Goldberg D, Gregory AM, et al. Depression and Generalized Anxiety Disorder: Cumulative and Sequential Comorbidity in a Birth Cohort Followed Prospectively to Age 32 Years. Archives of General Psychiatry. 2007 June 1, 2007;64(6):651–660.

[64] Silverstein B. Gender difference in the prevalence of clinical depression: the role played by depression associated with somatic symptoms. The American Journal of Psychiatry. 1999 March 1, 1999;156(3):480–482.

[65] Akiskal HS. Subaffective disorders: dysthymic, cyclothymic and bipolar II disorders in the "borderline" realm. Psychiatric Clinics of North America. 1981 Apr;4(1):25–46. PubMed PMID: 7232236. Epub 1981/04/01. eng.

[66] Cooke RG, Young LT, Levitt AJ, Pearce MM, Joffe RT. Bipolar II: Not so different when co-morbidity excluded. Depression. 1995;3(3):154–156.

[67] Dilsaver SC, Chen YW. Social phobia, panic disorder and suicidality in subjects with pure and depressive mania. Journal of Affective Disorders. 2003 Nov;77(2):173–177. PubMed PMID: 14607395. Epub 2003/11/11. eng.

[68] MacQueen GM, Marriott M, Begin H, Robb J, Joffe RT, Young LT. Subsyndromal symptoms assessed in longitudinal, prospective follow-up of a cohort of patients with bipolar disorder. Bipolar Disorders. 2003 Oct;5(5):349–355. PubMed PMID: 14525555. Epub 2003/10/04. eng.

[69] Cassano GB, Pini S, Saettoni M, Dell'Osso L. Multiple anxiety disorder comorbidity in patients with mood spectrum disorders with psychotic features. The American Journal of Psychiatry. 1999 Mar;156(3):474–476. PubMed PMID: 10080568. Epub 1999/03/18. eng.

[70] MacKinnon DF, Xu J, McMahon FJ, Simpson SG, Stine OC, McInnis MG, et al. Bipolar disorder and panic disorder in families: an analysis of chromosome 18 data. The American Journal of Psychiatry. 1998 Jun;155(6):829–381. PubMed PMID: 9619158. Epub 1998/06/10. eng.

[71] Judd LL, Akiskal HS, Schettler PJ, Coryell W, Endicott J, Maser JD, et al. A prospective investigation of the natural history of the long-term weekly symptomatic status of bipolar II disorder. Archives of General Psychiatry. 2003 Mar;60(3):261-269. PubMed PMID: 12622659. Epub 2003/03/08. eng.

[72] Kelsoe JR. Arguments for the genetic basis of the bipolar spectrum. Journal of Affective Disorders. 2003 Jan;73(1–2):183–97. PubMed PMID: 12507751. Epub 2003/01/01. eng.

[73] McMahon FJ, Simpson SG, McInnis MG, Badner JA, MacKinnon DF, DePaulo JR. Linkage of bipolar disorder to chromosome 18q and the validity of bipolar II disorder. Archives of General Psychiatry. 2001 Nov;58(11):1025–31. PubMed PMID: 11695948. Epub 2001/12/26. eng.

[74] Craddock N, Sklar P. Genetics of bipolar disorder: successful start to a long journey. Trends Genetics. 2009 Feb;25(2):99–105. PubMed PMID: 19144440. Epub 2009/01/16. eng.

[75] Kendler KS, Pedersen NL, Neale MC, Mathe AA. A pilot Swedish twin study of affective illness including hospital- and population-ascertained subsamples: results of model fitting. Behaviour Genetics. 1995 May;25(3):217–232. PubMed PMID: 7598665. Epub 1995/05/01. eng.

[76] McGuffin P, Rijsdijk F, Andrew M, Sham P, Katz R, Cardno A. The heritability of bipolar affective disorder and the genetic relationship to unipolar depression. Archives of General Psychiatry. 2003 May;60(5):497–502. PubMed PMID: 12742871. Epub 2003/05/14. eng.

[77] Coryell W, Endicott J, Reich T, Andreasen N, Keller M. A family study of bipolar II disorder. British Journal of Psychiatry. 1984 Jul;145:49–54. PubMed PMID: 6743943. Epub 1984/07/01. eng.

[78] Lee SY, Chen SL, Chen SH, Huang SY, Tzeng NS, Chang YH, et al. The COMT and DRD3 genes interacted in bipolar I but not bipolar II disorder. CORD Conference Proceedings. 2011;12(5):385–391. Eng.

[79] Sadovnick AD, Remick RA, Lam R, Zis AP, Yee IM, Huggins MJ, et al. Mood Disorder Service Genetic Database: morbidity risks for mood disorders in 3,942 first-degree relatives of 671 index cases with single depression, recurrent depression, bipolar I, or bipolar II. American Journal of Medical Genetics. 1994 Jun 15;54(2):132–140. PubMed PMID: 8074163. Epub 1994/06/15. eng.

[80] Schwartz JC, Diaz J, Pilon C, Sokoloff P. Possible implications of the dopamine D(3) receptor in schizophrenia and in antipsychotic drug actions. Brain Research. Brain Research Reviews. 2000 Mar;31(2–3):277–287. PubMed PMID: 10719154. Epub 2000/03/17. eng.

[81] Sokoloff P, Giros B, Martres M-P, Bouthenet M-L, Schwartz J-C. Molecular cloning and characterization of a novel dopamine receptor (D3) as a target for neuroleptics. Nature. 1990;347(6289):146–151.

[82] Jonsson EG, Nothen MM, Grunhage F, Farde L, Nakashima Y, Propping P, et al. Polymorphisms in the dopamine D2 receptor gene and their relationships to striatal dopamine receptor density of healthy volunteers. Molecular Psychiatry. 1999 May;4(3):290–296. PubMed PMID: 10395223. Epub 1999/07/08. eng.

[83] Kirov G, Jones I, McCandless F, Craddock N, Owen MJ. Family-based association studies of bipolar disorder with candidate genes involved in dopamine neurotransmission: DBH, DAT1, COMT, DRD2, DRD3 and DRD5. Molecular Psychiatry. 1999 Nov;4(6):558–565. PubMed PMID: 10578238. Epub 1999/12/01. eng.

[84] Souery D, Lipp O, Mahieu B, Mendelbaum K, De Martelaer V, Van Broeckhoven C, et al. Association study of bipolar disorder with candidate genes involved in catecholamine neurotransmission: DRD2, DRD3, DAT1, and TH genes. American Journal of Medical Genetics. 1996 Nov 22;67(6):551–555. PubMed PMID: 8950413. Epub 1996/11/22. eng.

[85] Lannfelt L, Sokoloff P, Martres M-P, Pilon C, Giros B, Jönsson E, et al. Amino acid substitution in the dopamine D3 receptor as a useful polymorphism for investigating psychiatric disorders. Psychiatric Genetics. 1992;2(4):249–256.

[86] Lundstrom K, Turpin MP. Proposed schizophrenia-related gene polymorphism: expression of the Ser9Gly mutant human dopamine D3 receptor with the Semliki Forest virus system. Biochemical and Biophysical Research Communication. 1996 Aug 23;225(3):1068–1072. PubMed PMID: 8780735. Epub 1996/08/23. eng.

[87] Scharfetter J, Chaudhry HR, Hornik K, Fuchs K, Sieghart W, Kasper S, et al. Dopamine D3 receptor gene polymorphism and response to clozapine in schizophrenic Pakastani patients. European Neuropsychopharmacology. 1999 Dec;10(1):17–20. PubMed PMID: 10647091. Epub 2000/01/26. eng.

[88] Shaikh S, Collier DA, Sham PC, Ball D, Aitchison K, Vallada H, et al. Allelic association between a Ser-9-Gly polymorphism in the dopamine D3 receptor gene and schizophrenia. Human Genetics. 1996 Jun;97(6):714–719. PubMed PMID: 8641685. Epub 1996/06/01. eng.

[89] Parsian A, Chakraverty S, Todd RD. Possible association between the dopamine D3 receptor gene and bipolar affective disorder. American Journal of Medicine Genetics. 1995 Jun 19;60(3):234–237. PubMed PMID: 7573178. Epub 1995/06/19. eng.

[90] Elvidge G, Jones I, McCandless F, Asherson P, Owen MJ, Craddock N. Allelic variation of a BalI polymorphism in the DRD3 gene does not influence susceptibility to bipolar disorder: results of analysis and meta-analysis. American Journal of Medicine Genetics. 2001 May 8;105(4):307–311. PubMed PMID: 11378841. Epub 2001/05/30. eng.

[91] Shaikh S, Ball D, Craddock N, Castle D, Hunt N, Mant R, et al. The dopamine D3 receptor gene: no association with bipolar affective disorder. Journal of Medical Genetics. 1993 April 1;30(4):308–309.

[92] Quraishi S, Frangou S. Neuropsychology of bipolar disorder: a review. Journal of Affective Disorders. 2002 Dec;72(3):209–226. PubMed PMID: 12450638. Epub 2002/11/27. eng.

[93] Baquet ZC, Bickford PC, Jones KR. Brain-derived neurotrophic factor is required for the establishment of the proper number of dopaminergic neurons in the substantia nigra pars compacta. Journal of Neuroscience. 2005 Jun 29;25(26):6251–6259. PubMed PMID: 15987955. Epub 2005/07/01. eng.

[94] Guillin O, Diaz J, Carroll P, Griffon N, Schwartz JC, Sokoloff P. BDNF controls dopamine D3 receptor expression and triggers behavioural sensitization. Nature. 2001 May 3;411(6833):86–89. PubMed PMID: 11333982. Epub 2001/05/03. eng.

[95] Chen ZY, Ieraci A, Teng H, Dall H, Meng CX, Herrera DG, et al. Sortilin controls intracellular sorting of brain-derived neurotrophic factor to the regulated secretory pathway. Journal of Neuroscience. 2005 Jun 29;25(26):6156–6166. PubMed PMID: 15987945. Pubmed Central PMCID: 1201519. Epub 2005/07/01. eng.

[96] Chen ZY, Jing D, Bath KG, Ieraci A, Khan T, Siao CJ, et al. Genetic variant BDNF (Val66Met) polymorphism alters anxiety-related behavior. Science. 2006 Oct 6;314(5796):140–143. PubMed PMID: 17023662. Pubmed Central PMCID: 1880880. Epub 2006/10/07. eng.

[97] Chen ZY, Patel PD, Sant G, Meng CX, Teng KK, Hempstead BL, et al. Variant brain-derived neurotrophic factor (BDNF) (Met66) alters the intracellular trafficking and activity-dependent secretion of wild-type BDNF in neurosecretory cells and cortical neurons. Journal of Neuroscience. 2004 May 5;24(18):4401–4411. PubMed PMID: 15128854. Epub 2004/05/07. eng.

[98] Egan MF, Kojima M, Callicott JH, Goldberg TE, Kolachana BS, Bertolino A, et al. The BDNF val66met polymorphism affects activity-dependent secretion of BDNF and human memory and hippocampal function. Cell. 2003 Jan 24;112(2):257–269. PubMed PMID: 12553913. Epub 2003/01/30. eng.

[99] Hong CJ, Huo SJ, Yen FC, Tung CL, Pan GM, Tsai SJ. Association study of a brain-derived neurotrophic-factor genetic polymorphism and mood disorders, age of onset and suicidal behavior. Neuropsychobiology. 2003;48(4):186–189. PubMed PMID: 14673216. Epub 2003/12/16. eng.

[100] Nakata K, Ujike H, Sakai A, Uchida N, Nomura A, Imamura T, et al. Association study of the brain-derived neurotrophic factor (BDNF) gene with bipolar disorder. Neuroscience Letter. 2003 Jan 30;337(1):17–20. PubMed PMID: 12524161. Epub 2003/01/14. eng.

[101] Tang J, Xiao L, Shu C, Wang G, Liu Z, Wang X, et al. Association of the brain-derived neurotrophic factor gene and bipolar disorder with early age of onset in mainland China. Neuroscience Letter. 2008 Mar 12;433(2):98–102. PubMed PMID: 18242852. Epub 2008/02/05. eng.

[102] Pivac N, Kim B, Nedic G, Joo YH, Kozaric-Kovacic D, Hong JP, et al. Ethnic differences in brain-derived neurotrophic factor Val66Met polymorphism in Croatian and Korean healthy participants. Croatian Medical Journal. 2009 Feb;50(1):43-48. PubMed PMID: 19260143. Pubmed Central PMCID: 2657559. Epub 2009/03/05. eng.

[103] Harris EL, Noyes R, Jr., Crowe RR, Chaudhry DR. Family study of agoraphobia. Report of a pilot study. Archives of General Psychiatry. 1983 Oct;40(10):1061–1064. PubMed PMID: 6625854.

[104] Knutson, B., Wolkowitz, O.M., Cole, S.W., Chan, T., Moore, E.A., Johnson, R.C., Terpstra, J., Turner, R.A., Reus, V.I. (1998). Selective alteration of personality and social behavior by serotonergic intervention. The American journal of psychiatry 155, 373-379.

[105] Wiers RW, Sergeant JA, Gunning WB. Psychological mechanisms of enhanced risk of addiction in children of alcoholics: a dual pathway? Acta Paediatrica. 1994 Nov;404:9–13. PubMed PMID: 7841640.

[106] McGue M. A behavioral-genetic perspective on children of alcoholics. Alcohol Health and Research World. 1997;21(3):210–217. PubMed PMID: 15706771.

[107] Ritchie T, Noble EP. Association of seven polymorphisms of the D2 dopamine receptor gene with brain receptor-binding characteristics. Neurochemical Research. 2003 Jan;28(1):73–82. PubMed PMID: 12587665. Epub 2003/02/18. eng.

[108] Chen CC, Lu RB, Chen YC, Wang MF, Chang YC, Li TK, et al. Interaction between the functional polymorphisms of the alcohol-metabolism genes in protection against alcoholism. The American Journal of Human Genetics. 1999 Sep;65(3):795–807. PubMed PMID: 10441588. Pubmed Central PMCID: 1377988. Epub 1999/08/12. eng.

[109] Agarwal DP, Goedde HW. Medicobiological and genetic studies on alcoholism. Role of metabolic variation and ethnicity on drinking habits, alcohol abuse and alcohol-related mortality. Journal of Clinical Investigation. 1992 Jun;70(6):465–477. PubMed PMID: 1392414. Epub 1992/06/01. eng.

[110] Thomasson HR, Edenberg HJ, Crabb DW, Mai XL, Jerome RE, Li TK, et al. Alcohol and aldehyde dehydrogenase genotypes and alcoholism in Chinese men. The American Journal of Human Genetics. 1991 Apr;48(4):677–681. PubMed PMID: 2014795. Pubmed Central PMCID: 1682953. Epub 1991/04/01. eng.

[111] Wang YS, Lee SY, Chen SL, Chang YH, Wang TY, Lin SH, et al. Role of DRD2 and ALDH2 genes in bipolar II disorder with and without comorbid anxiety disorder. European Psychiatry: The Journal of the Association of European Psychiatrists. 2014 Mar;29(3):142–148. PubMed PMID: 23835015. Epub 2013/07/10. eng.

[112] Doughty CJ, Wells JE, Joyce PR, Olds RJ, Walsh AE. Bipolar-panic disorder comorbidity within bipolar disorder families: a study of siblings. Bipolar Disorders. 2004 Jun;6(3):245–252. PubMed PMID: 15117403. Epub 2004/05/01. eng.

[113] Rotondo A, Mazzanti C, Dell'Osso L, Rucci P, Sullivan P, Bouanani S, et al. Catechol o-methyltransferase, serotonin transporter, and tryptophan hydroxylase gene polymorphisms in bipolar disorder patients with and without comorbid panic disorder. The American Journal of Psychiatry. 2002 Jan;159(1):23–29. PubMed PMID: 11772685. Epub 2002/01/05. eng.

[114] Chang YH, Lee SY, Chen SL, Tzeng NS, Wang TY, Lee IH, et al. Genetic variants of the BDNF and DRD3 genes in bipolar disorder comorbid with anxiety disorder. Journal of Affective Disorders. 2013 Dec;151(3):967–972. PubMed PMID: 24021960. Epub 2013/09/12. eng.

[115] Joyce PR, Rogers GR, Miller AL, Mulder RT, Luty SE, Kennedy MA. Polymorphisms of DRD4 and DRD3 and risk of avoidant and obsessive personality traits and disorders. Psychiatry Research. 2003 Jul 15;119(1–2):1–10. PubMed PMID: 12860355. Epub 2003/07/16. eng.

[116] Light KJ, Joyce PR, Luty SE, Mulder RT, Frampton CMA, Joyce LRM, et al. Preliminary evidence for an association between a dopamine D3 receptor gene variant and obsessive-

compulsive personality disorder in patients with major depression. American Journal of Medical Genetics Part B: Neuropsychiatric Genetics. 2006;141B(4):409–413.

[117] Takahashi T, Suzuki M, Tsunoda M, Kawamura Y, Takahashi N, Maeno N, et al. The association of genotypic combination of the DRD3 and BDNF polymorphisms on the adhesion interthalamica and medial temporal lobe structures. Progress in Neuro-Psychopharmacology & Biological Psychiatry. 2008 Jul 1;32(5):1236–1242. PubMed PMID: 18472202. Epub 2008/05/13. eng.

[118] Gourion D, Goldberger C, Leroy S, Bourdel M-C, Olié J-P, Krebs M-O. Age at onset of schizophrenia: interaction between brain-derived neurotrophic factor and dopamine D3 receptor gene variants. NeuroReport. 2005;16(12):1407–1410.

[119] Chang CT, Chang YH, Yung-Wei Wu J, Lee SY, Chen SL, Chen SH, et al. Neuropsychological functions impairment in different subtypes of bipolar disorder with or without comorbid anxiety disorders. Psychiatry Research. 2012 Jun 27 ; 200(2-3):246-51. PubMed PMID: 22748188. Epub 2012/07/04. Eng.

[120] Lohoff FW, Sander T, Ferraro TN, Dahl JP, Gallinat J, Berrettini WH. Confirmation of association between the Val66Met polymorphism in the brain-derived neurotrophic factor (BDNF) gene and bipolar I disorder. American Journal of Medical Genetics Part B: Neuropsychiatric Genetics. 2005;139B(1):51–53.

[121] Frustaci A, Pozzi G, Gianfagna F, Manzoli L, Boccia S. Meta-analysis of the brain-derived neurotrophic factor gene (BDNF) Val66Met polymorphism in anxiety disorders and anxiety-related personality traits. Neuropsychobiology. 2008;58(3–4):163–170. PubMed PMID: 19088493. Epub 2008/12/18. eng.

[122] Jiang X, Xu K, Hoberman J, Tian F, Marko AJ, Waheed JF, et al. BDNF variation and mood disorders: a novel functional promoter polymorphism and Val66Met are associated with anxiety but have opposing effects. Neuropsychopharmacology. 2005 Jul;30(7):1353–1361. PubMed PMID: 15770238. Epub 2005/03/17. eng.

[123] Driessen M, Meier S, Hill A, Wetterling T, Lange W, Junghanns K. The course of anxiety, depression and drinking behaviours after completed detoxification in alcoholics with and without comorbid anxiety and depressive disorders. Alcohol and Alcoholism (Oxford, Oxfordshire). 2001 May–Jun;36(3):249–255. PubMed PMID: 11373263. Epub 2001/05/25. eng.

[124] Finckh U, Rommelspacher H, Kuhn S, Dufeu P, Otto G, Heinz A, et al. Influence of the dopamine D2 receptor (DRD2) genotype on neuroadaptive effects of alcohol and the clinical outcome of alcoholism. Pharmacogenetics. 1997 Aug;7(4):271–281. PubMed PMID: 9295055.

[125] Lucht MJ, Kuehn KU, Schroeder W, Armbruster J, Abraham G, Schattenberg A, et al. Influence of the dopamine D2 receptor (DRD2) exon 8 genotype on efficacy of tiapride and clinical outcome of alcohol withdrawal. Pharmacogenetics. 2001 Nov;11(8):647–653. PubMed PMID: 11692072.

http://dx.doi.org/10.5772/66117

[126] Lawford BR, Young RM, Rowell JA, Qualichefski J, Fletcher BH, Syndulko K, et al. Bromocriptine in the treatment of alcoholics with the D2 dopamine receptor A1 allele. Nature Medicine. 1995 Apr;1(4):337–341. PubMed PMID: 7585063.

[127] Huang SY, Lin WW, Ko HC, Lee JF, Wang TJ, Chou YH, et al. Possible interaction of alcohol dehydrogenase and aldehyde dehydrogenase genes with the dopamine D2 receptor gene in anxiety-depressive alcohol dependence. Alcoholism, Clinical and Experimental Research. 2004 Mar;28(3):374–384. PubMed PMID: 15084894.

[128] Huang, S.Y., Lin, W.W., Wan, F.J., Chang, A.J., Ko, H.C., Wang, T.J., Wu, P.L., Lu, R.B. (2007). Monoamine oxidase-A polymorphisms might modify the association between the dopamine D2 receptor gene and alcohol dependence. Journal of psychiatry & neuroscience : JPN 32, 185-192.

[129] Dikeos DG, Papadimitriou GN, Avramopoulos D, Karadima G, Daskalopoulou EG, Souery D, et al. Association between the dopamine D3 receptor gene locus (DRD3) and unipolar affective disorder. Psychiatr Genet. 1999;9(4):189-95.

[130] Neria Y, Besser A, Kiper D, Westphal M. A longitudinal study of posttraumatic stress disorder, depression, and generalized anxiety disorder in Israeli civilians exposed to war trauma. J Trauma Stress. 2010;23(3):322-30.

[131] Judd LL, Akiskal HS, Schettler PJ, Coryell W, Maser J, Rice JA, et al. The comparative clinical phenotype and long term longitudinal episode course of bipolar I and II: a clinical spectrum or distinct disorders? Journal of affective disorders. 2003;73(1-2):19-32.

[132] Pallanti S, Quercioli L, Pazzagli A, Rossi A, Dell'Osso L, Pini S, et al. Awareness of illness and subjective experience of cognitive complaints in patients with bipolar I and bipolar II disorder. The American journal of psychiatry. 1999;156(7):1094-6.

[133] Vieta E, Gasto C, Otero A, Nieto E, Vallejo J. Differential features between bipolar I and bipolar II disorder. Comprehensive psychiatry. 1997;38(2):98-101.

[134] Martinowich K, Lu B. Interaction between BDNF and serotonin: role in mood disorders. Neuropsychopharmacology: Official Publication of the American College of Neuropsychopharmacology. 2008;33(1):73–83.

[135] Mossner R, Daniel S, Albert D, Heils A, Okladnova O, Schmitt A, et al. Serotonin transporter function is modulated by brain-derived neurotrophic factor (BDNF) but not nerve growth factor (NGF). Neurochemistry International. 2000 Mar;36(3):197–202. PubMed PMID: 10676853. Epub 2000/02/17. eng.

Chapter 5

Complex Comorbidity of Substance Use Disorders with Anxiety Disorders: Diagnosis and Treatment

Onat Yilmaz and Nesrin Dilbaz

Additional information is available at the end of the chapter

http://dx.doi.org/10.5772/66247

Abstract

Substance use disorders is a worldwide public health problem that commonly ocur together with our psychiatric and medical disorders. Along with etiologic origins, prognosis of anxiety disorders intercepts with substance use disorders. Due to overlapping symptoms and complaints, it is always difficult for clinicians to recognise these disorders separately. In addition, selecting the best treatment approach is challenging because of the relative risk for developing anxiety disorders in substance use patients or vice versa. In this chapter, authors are focused on adding new aspects to the clinicians for evaluating, treating and following patients with comorbid substance use disorder and anxiety disorder.

Keywords: anxiety disorder, substance use disorder, comorbidity, diagnosis, treatment

1. Introduction

Comorbidity is a medical term, implying any distinct additional clinic condition that exists during the progress of an index disease in a patient. Growing evidence has revealed various associations or interactions between many mental disorders. Nevertheless, when evolving diagnostic criteria and symptom diversity are taken into account, researchers developed different definitions in order to build a consensus among clinicians for patients meeting criteria for more than one mental disorder. Several authors asserted the terms "dual diagnosis", "concurrent disorders" and "co-occurring disorders" for defining such patients, and the authors of this chapter will use the term "comorbidity" to indicate any anxiety disorder and any substance use disorder (SUD) co-observed in the same individual. Although there is insufficient data about the underlying mechanisms of psychiatric comorbidity, the presence of comorbidity is generally associated with worsened prognosis, poor treatment outcomes, medication abuse risk and lower treatment compliance.

As a statistical rule, disorders with high prevalence rates might be expected to be diagnosed in the same individual concurrently. When it comes to mental disorders, epidemiological studies in general population have detected a frequent occurrence pattern. Taken together, despite some methodological issues, anxiety disorders and substance use disorders are both disabling conditions, which often occur together.

2. Epidemiology

Data from clinical studies and community-based studies report a significant association between these disorders. In their first article derived from the results of the National Institute on Alcohol Abuse and Alcoholism's National Epidemiologic Survey on Alcohol and Related Conditions (NESARC), Grant et al. used Diagnostic and Statistical Manual of Mental Disorders, Fourth Edition (DSM–IV) definitions for this study and found that 12-month prevalence of any anxiety disorder was 11.08%, 12-month prevalence of any SUDs was 9.35% [1]. Among individuals with any SUDs, 17.71% had at least one independent anxiety disorder during the last 12-month period, and 1.46–10.54% had a specific anxiety disorder. Among participants with an anxiety disorders, 14.96% of them had at least one substance use disorder. Diagnosed with alcohol and drug use disorders increased the risk for any anxiety disorders 1.7–2.8 times. A few years later Grant et al. announced weighted 1-year incidence rates of any anxiety disorders and any drug use disorders, as 1.57 and 0.31, respectively [2]. In their NESARC-III study, in which they used DSM-5 criteria, they found prevalence rates of 12-month any drug use disorder as 3.9% and rates of lifetime any drug use disorder as 9.9%. Also, they announced adjusted odd ratios for the association between 12-month drug use disorder and any anxiety disorder as 1.2 and 1.3 for the association between lifetime drug use disorder and any anxiety disorder [3]. Using data from NESARC study, Conway et al. postulated that lifetime prevalence rates of any anxiety disorder among people diagnosed with drug use disorders are higher than the general population, 29.9 versus 16.16% [4]. Compton et al. used the same data and announced that after adjusting for psychiatric comorbidity, odds ratios detected statistically significant only between any anxiety disorder and drug dependence [5]. According to Mericle et al., approximately 8.2% of white Americans, 5.8% of Latin Americans, and 5.4% of Black Americans 2.1% of Asian Americans met criteria for lifetime co-occurring substance use and mental disorders. With regard to the data composed of three different community-based surveys, Mericle et al. suggested that 55.4% of individuals diagnosed with any SUDs also met criteria for any mental disorders, and among them, any anxiety disorders were the highest with a co-occurance rate of 47.3% [6].

Community-based epidemiologic studies have also been conducted in other countries. In their latest study, derived from the data from the 2007 National Survey of Mental Health and Wellbeing, McEvoy et al. found no significant relationship between any anxiety disorder and any SUD [7]. Merikangas et al. analysed six studies from Germany, Mexico, Netherlands, United States and Canada and found a strong association between anxiety disorders and SUDs. Comorbidity rates of anxiety disorders and SUDs were different among these countries and varied between 9.9 and 56% [8]. Leray et al. announced that overall prevalence of

http://dx.doi.org/10.5772/66247

Study	Sample	Diagnostic tools	Results
Grant et al. 2006 [1]	43,093 Americans NESARC	DSM-IV (AUDADIS)	In a 12-month period 17.71% of individuals with SUD met criteria for AD
Grant et al. [2]	34,653 Americans NESARC	DSM-IV (AUDADIS)	Weighted incidence for any AD is 1.57, for any SUD is 0.31. Low risk for comorbidity
Grant et al. [3]	36,309 Americans NESARC	DSM-5 (AUDADIS)	People with DUD are 1.2 times of greater risk of having any AD in 12-month period, 1.3 times of greater risk for lifetime
Compton et al. [5]	43,093 Americans NESARC	DSM-IV (AUDADIS)	Among all mental disorders in the study, odds ratios are only significant between any AD and drug dependence
Conway et al. [4]	43,093 Americans NESARC	DSM-IV (AUDADIS)	Prevalence of any AD is higher among individuals with DUD than the normal population, 29.9 vs 16.16%
Mericle et al. [6]	20,013 US Citizens CPES	DSM-IV (WMH-CIDI)	Prevalence rate of any AD among people with SUD is 47.3%
McEvoy et al. [7]	8841 Australians NSMHWB	DSM-IV (WMH-CIDI)	No significant association between SUD and any AD
Merikangas et al. [8]	USA, n = 2874 Germany, n = 3021 Mexico, n = 1734 Netherlands, n = 7076 Ontario, n = 6902 USA, n = 8098 ICPE	DSM-III-R	Different comorbidity rates between AD and SUD, ranging from 9.9 to 56%
Leray et al. [9]	36,105 French citizens MHGP	MINI	3.7% of people with any AD met criteria for drug addiction
Toftdahl et al. [10]	463,003 Dutch citizens	ICD-10	Prevalence rate of any SUDs in patients with AD, PTSD and OCD is 24.8, 17 and 11.4%, respectively
Lai et al. [11]	22 epidemiologic surveys	Variable diagnostic methods	People diagnosed with any SUD are of 2.9 times greater risk of diagnosing any AD

Only studies with epidemiologic surveys are included. NESARC, National Epidemiological Survey of Alcohol and Related Conditions; DSM-IV, Diagnostic and Statistical Manual of Mental Disorders, Fourth Edition; AUDADIS, National Institute on Alcohol Abuse and Alcoholism's Alcohol Use Disorder and Associated Disabilities Interview Schedule; AD, anxiety disorder; SUD, substance use disorder; DUD, drug use disorder; DSM-5, Diagnostic and Statistical Manual of Mental Disorders, Fifth Edition; CPES, Collaborative Psychiatric Epidemiology Studies; NSMHWB, National Survey on Mental Health and Wellbeing; WMH-CIDI, World Mental Health-Composite International Diagnostic Interview; ICPE, International Consortium in Psychiatric Epidemiology; DSM-III-R, Diagnostic and Statistical Manual of Mental Disorders, Third Edition, Revised Form; OCD, obsessive-compulsive disorder; PTSD, posttraumatic stress disorder.

Table 1. Brief information about recent population-based studies of anxiety disorders-SUD comorbidity.

anxiety disorders in adult French population was estimated to be 21.6 and 3.7% of individuals with a diagnosis of an anxiety disorders also met criteria for drug addiction [9]. In a recent population-based study from Denmark, Toftdahl et al. established prevalence rates of SUDs in patients with anxiety disorders, posttraumatic stress disorder (PTSD) and obsessive-compulsive disorders (OCD) as 24.8, 17 and 11.4%, respectively [10]. In a recent meta-analyses and systematic review in which 22 community-based studies from different countries were evaluated, Lai et al. found high association between illicit drug use and any anxiety disorder (pooled odds ratio 2.907). In other words, people with SUDs were 2.9 times of greater risk of having any anxiety disorder [11]. **Table 1** presents the brief information about the studies aforementioned.

Epidemiologic studies focusing on clinical cases often examined prevalence of anxiety disorders (ADs) in individuals with SUD and prevalence estimates range from 4 to 80.3%. In their prospective cohort study, Franken and Hendriks detected lifetime prevalence of any AD as

Study	**Sample**	**Diagnostic tools**	**Results**
Franken and Hendriks [12]	116 inpatients under substance abuse treatment	CIDI SCL-90 DSM-III-R ASI-Drug	Lifetime prevalence of any AD was 53.4%, current prevalence was 38.8%
Bakken et al. [13]	146 alcohol-dependent inpatients 114 poly-substance-dependent inpatients	DSM-IV CIDI	Weighted incidence for any AD is 1.57, for any SUD is 0.31. Low risk for comorbidity
Bakken et al. [14]	60 poly-substance abusers, 194 alcohol-dependent patients	DSM-IV CIDI HSCL-25	48.6% of the sample met criteria for SAD
Sbrana et al. [15]	287 patients with various mental disorders, inpatient and outpatient setting	SCID	4% of patients with OCD also had DUD, 6% of the patients with PD also had DUD
Rosen et al. [16]	140 opiat-dependent patients over 50 years old	CIDI SF-12v2	27.8% of dependent patients also had PTSD, 29.7% of dependent patients had GAD
Smith and Book [17]	56 outpatients involved in substance abuse treatment program	SCID PSWQ ASI BDI	Prevalence rate of any AD among people with SUD is 47.3%
Goldner et al. [18]	Systematic review and meta-analysis, including 11 studies	Various diagnostic tools	29% of the patients in the studies met criteria for ADs and 50% of the participants reported anxiety symptoms

Only studies consisting of clinical samples are included. DSM-III-R, Diagnostic and Statistical Manual of Mental Disorders, Third Edition, Revised Form; ASI-Drug, ASI Drug Use Severity Composite Score; CIDI, Composite International Diagnostic Interview; SCL-90, Symptom Checklist-90 Revised; DSM-IV, Diagnostic and Statistical Manual of Mental Disorders, Fourth Edition; HSCL-25, Hopkins Symptom Checklist; SAD, social anxiety disorder; SCID, Structured Clinical Interview for DSM-IV; PD, panic disorder; DUD, drug use disorder; SF-12v2, Short form health survey; PTSD, posttraumatic stress disorder; GAD, generalised anxiety disorder; ASI, anxiety sensitivity index; BDI, Beck depression Inventory; PSWQ, Penn State Worry Questionnaire.

Table 2. Brief information about recent clinical studies of anxiety disorders-SUD comorbidity.

53.4% and current prevalence of AD as 38.8% in an inpatient substance abuse population [12]. Bakken et al. asserted that current social anxiety disorder (SAD) diagnosis among patients with SUD was 51%, and this rate was higher than the alcohol-dependent patient group [13]. In their later study, Bakken et al. reported 48.6% of their study population met criteria for SAD [14]. Sbrana et al. evaluated both inpatients and outpatients diagnosed with mental disorders and found that 4% of the participants with OCD had drug use disorder, and 6% of the participants with panic disorder (PD) had drug use disorder [15]. In patients with opiate dependency, Rosen et al. found the prevalence of PTSD and generalised anxiety disorder (GAD) as 27.8 and 29.7%, respectively [16]. Smith and Book analysed 56 patients with substance use disorder and postulated that 32% of the patient group met criteria for GAD [17]. Goldner et al. made a systematic review about patients with non-medical prescription opioid users and reported that 29% of the patients in the studies had the diagnosis of AD, and 50% of the patients were reporting anxiety symptoms [18]. **Table 2** presents the brief information about these studies.

Current literature indicates a comorbidity between substance use disorders and anxiety disorders; however, while evaluating the study results, clinicians should pay attention to some important issues including study designs, population or participants included in the study, diagnostic tools or criteria used, population size. Individuals with SUD involved in the studies may be in different stages of the disorder: experiencing withdrawal symptoms or abstinent, etc. Current diagnostic instruments are insufficiently sensitive to discriminate AD symptoms or substance-related symptoms, and training of the interviewers participated in the study for the diagnostic tools might not be standardised.

3. Development and maintenance of substance use-anxiety comorbidity

Based on search results, clinical experience and expert opinions, these disorders are both in a causal and etiologic relationship in both the development and maintenance of this comorbidity. Different models have been developed in order to shed light to the comorbidity of anxiety and substance use disorders. In direct causal model, a primary psychopathology causes a secondary psychopathology. It has been well known that some substances including alcohol reduce anxiety and stress levels in a short time after they are consumed. This model is named as "tension-reduction" [19], "self-medication" [20] and "stress-response-dampening" [21] by different researchers. In all models, individuals use substances in order to reduce their anxiety levels and therefore, in most of the studies evaluating AD-SUD comorbidity, results indicate AD precedes the SUD. Smith and Book found GAD preceding to SUD in their study [17]. Having PTSD diagnosis was a risk factor for SUD, reported in a 10-year longitudinal study [22]. But also, according to the vulnerability hypothesis, any substance use increases anxiety levels and psychological arousal in time and makes the individual prone to develop PTSD after traumatic event or stress. Robinson et al. found evidence supporting self-medication theory and put forward the idea that self-medication in AD constitutes risk for developing SUD or vice versa [23]. Swendsen and Conway found that except GAD, ADs were predictors of later drug dependence [24]. In another community-based study, SAD was found to be a risk factor for cannabis dependence [25]. On the contrary, there are studies pretending that substance-related symptoms and/or substance withdrawal symptoms might play a role in the onset of AD. Using the data from a large community sample, authors claimed that lifetime cannabis use was found to be associated with lifetime and current panic disorder (PD) diagnosis, but current cannabis use was associated with only panic attacks [26]. Brady et al. conducted a clinic study and asserted that half of the patients had PTSD before onset of cocaine dependence, and the other half met criteria for cocaine dependence before PTSD diagnosis [27]. According to Stewart and Conrod, once the comorbidity is present, these disorders pretend to serve to maintain the existence of one another. In bidirectional model, an individual with an AD may use substances in order to relieve symptoms of anxiety or stress. In time, after SUD develops, experiencing craving or withdrawal might exacerbate physical symptoms, and thus, panic attacks may be triggered or anticipation anxiety for the attacks may escalate [28].

Although epidemiological evidences are relatively determined, there is subtle data in the literature regarding biological etiology of this comorbidity. Hodgson et al. examined a sample of Mexican-Americans and identified a region on chromosome 18, possibly responsible for drug dependence and anxiety comorbidity [29].

4. Assessment and diagnosis

In the literature, assessment of comorbid mental disorders is generally performed in non-SUD psychiatric patients, and it is a bit more complicated to assess the comorbid mental disorders of patients with current diagnosis of SUD. In individuals with SUD who are applying to health care providers, it might be difficult for professionals to distinguish anxiety symptoms from the anxiety disorder. When evaluating the presenting symptoms, it is better to keep in mind that symptoms might be independent of the substance used, might be secondary to intoxication or withdrawal or might be due to an unidentified effect of the substance used. Diagnosing the comorbidity is often complicated by the anxiolytic or anxiogenic effects of the substance used. Being an extension of the primary/secondary disorder model, the DSM-IV and DSM-V classifies "substance-induced disorders (SIDs)" as a distinct diagnostic entity, which occurs during a period of heavy substance use or within the first 4 weeks of withdrawal. To diagnose a SID, symptoms should be heavier and more disabling than the symptoms expected to be presented in withdrawal or craving. A question may arise while evaluating a patient with SUD. Would diagnosing an AD in a patient with SUD require great distress and social/occupational impairment or would endorsement of the symptoms/diagnostic criteria by the patient be enough for the diagnosis?

Structured Clinical Interview for DSM-V (SCID) is a semi-structured interview tool, designed for clinicians and interviewers with adequate training and/or clinical expertise. Having four different versions (clinician, personality disorders, research and clinical trials), it is offered to be used after a short overview and general screening of the patient. Although administration of the tool can take up to several hours, test-retest study of DSM-IV diagnoses postulated excellent reliability of Axis I and Axis II diagnoses and SUDs [30]. Because established questionnaires and scales implemented in non-abusers might misdirect both the patient and the clinician. In line with this hypothesis, Kranzler et al. conducted a research in a population with SUD and asserted that, SCID had poor validity for AD in such populations [31]. Kranzler et al. also asked the efficiency of the clinicians in diagnosing psychiatric disorders in patients with SUDs and found that clinicians using an unstructured clinical interview may be effective in diagnosing SUD, but, on the other hand, they may fail in diagnosing comorbid psychiatric disorder. They offered using SCID by trained clinicians may enhance the validity of both SUD diagnosis and comorbid disorders [32].

Thus, a more reliable and valid diagnostic tool for evaluating psychiatric disorders among substance abusers, Hasin and Grant developed the semi-structured diagnostic interview, Psychiatric Research Interview for Substance and Mental Disorders (PRISM) [33]. As well as substance use diagnosis, clinicians can use PRISM in order to evaluate current or lifetime diagnosis of some psychiatric disorders. Diagnostic sections of PRISM are modular, providing the interviewer to use it for treatment or research needs. Question patterns in the substance use module need gathering further and detailed information from the patient, if only the response is "yes" to any of them. After publication of DSM-IV, PRISM adapted to PRISM-IV, but one study found that the reliability of some DSM-IV ADs in the adapted form was lower when compared to other mental disorders [34].

Developed by the National Institute on Alcohol Abuse and Alcoholism (NIAAA), The Alcohol Use Disorder and Associated Disabilities Interview Schedule (AUDADIS) is used for diagnosing current or lifetime diagnoses of mood, anxiety, SUD and personality disorders. Although it is commonly used for population-based surveys, clinicians can use AUDADIS to tailor the treatment of the patient based on the responses to this schedule. Hasin et al. found this semi-structured interview concordant with PRISM-5 and suggested them for determining dimensional measures of psychopathology [35]. Developers of the tool recommend using computer-assisted version.

The Composite International Diagnostic Interview (CIDI), developed by World Health Organisation, helps clinicians in gathering information both for diagnosis and treatment planning. Concordance of CIDI with SCID found to be excellent for alcohol dependence, fair for drug dependence, good for alcohol and drug abuse [36].

The assessment of patients with AD and SUD comorbidity is a sensitive process, which requires a good therapeutic relationship, detailed medical history and both physical and psychiatric examination. Family history of any mental disorders should be interviewed and establishing contact with a spouse or a family member would bring contribution. Drawing a timeline representing the course of the psychiatric and physical symptoms and diagnosis would be helpful in detecting periods of exacerbation and remission.

Evaluating the patient during period of abstinence from substances is generally recommended, but it can be difficult in clinical practice. Forming a retrospective timeline showing both anxiety and substance use symptoms longitudinally might shed light into temporal sequence and the presence of anxiety symptoms during abstinence periods. Patient might create a prospective diary system by recording substance use and physical/mental symptoms. This system might help understanding which situations or emotions direct the individual to substance use and whether withdrawal or craving can precipitate or exacerbate symptoms of anxiety. The information gathered from this diary system would also be useful for the clinician in formulating a treatment plan and later evaluation of the efficacy of the treatment interventions [37].

5. Prognosis in comorbidity

Recent evidence indicates current diagnosis of AD result in poorer treatment results and increased relapse risk in patients with SUD. In other words, patients who have anxious complaints or symptoms during or after SUD treatment might carry risks for using various substances to alleviate anxiety symptoms. As Smith and Book reported in their study, GAD interferes with SUD treatment [17]. In Oumette et al.'s follow-up study, comorbid AD impaired engagement in SUD treatment, and comorbid AD-SUD patients had worse results in functioning and symptom measures when compared with patients with SUD alone. About 41.6% of patients with comorbid diagnosis reported significant distress [38]. Ford et al. reported that severity of PTSD symptoms is associated with poor contingency management, but not standard treatment [39]. In their follow-up study, Mills et al. found no relation between PTSD and poor treatment initiation/completion of SUD. But also in this study, 2 years after the beginning of treatment, PTSD symptoms remained significant and caused disability [40]. In addition to these results, Tomasson and Vaglum found better SUD treatment outcomes with comorbid ADs [41].

Brown et al. found that in women with PTSD, relapse occurred more rapidly than non-PTSD women and the authors claimed that substance using individuals with PTSD return immediately to using substances to relieve their anxiety symptoms [42]. Brown also examined 6-month outcomes of women with PTSD-SUD diagnosis, and relapse was observed at 52% of the patients. Twenty-four percent of the population remitted from PTSD. In this study, only significant predictor of substance use relapse was the severity of PTSD re-experiencing symptoms at attendance. Re-experiencing symptoms were also identified as predictor of PTSD status at follow-up [43]. Oumette et al. evaluated male patients in their post-treatment period for 2 years and revealed that SUD-PTSD individuals had poorer treatment outcomes, and also SUD-only individuals were more likely to have remitted than SUD-PTSD group [44]. In another study by Oumette et al., they reported that attendance to the 12-Step program was associated with remission (defined as abstinence from all drugs, no problems associated with substance use and minimal alcohol use). They also emphasised that soon after discharge from acute SUD treatment, starting PTSD-oriented treatment was predictor of SUD remission 5 years later [45, 46]. Taken together, PTSD-SUD individuals tend to relapse more quickly following SUD treatment, PTSD re-experiencing symptom severity is a predictor of relapse to substance use, starting the PTSD treatment soon after SUD treatment is associated with better substance use outcomes in 5 years.

There is limited information about PD-SUD comorbidity in the literature. In a community-based study, among other mental disorders, PD had the greatest odd ratio for being diagnosed with current SUD [47]. Fals-Stewart and Schafer examined the effects of OCD diagnosis

http://dx.doi.org/10.5772/66247

on SUD, and among participants, those who took interventions for OCD had better treatment outcomes and had longer periods of abstinence [48]. According to Bruce et al., the presence of a SUD decreased the treatment outcomes of GAD by nearly fivefold and increased the recurrence of GAD symptoms by nearly threefold. There were no differences in treatment responses or recurrence rates for social phobia at 12 year follow-up period [49]. Differences in the results of the studies may be related to diagnostic instruments, nature of the substance used and duration of treatment administered.

In their systematic review, Fatseas et al. evaluated 12 studies available from the literature and found lifetime prevalence of anxiety disorders ranging from 26 to 35% in opiate-dependent patients under treatment. In opiate-dependent patients anxiety comorbidity had a complex and heterogeneous aetiology, but the current data recommend treating anxiety disorders in the first place in order to prevent opiate dependence. Among ADs in these studies, phobic disorders often precede the onset of opiate-dependence [50].

6. Treatment approaches

Even SUD-AD comorbidity has unfavourable outcomes, patients and families often encounter economic and systemic barriers to achieve treatment. Between 2003 and 2006, only 9–11% of those who could benefit from drug treatment could able to receive treatment [51]. In a review, current barriers to reach treatment were defined as payer financing systems, clinical and organisational limitations and confidentiality of patient records [52].

The most important part of treatment planning is deciding when (immediate or elective) and how (pharmacoterapy, psychotherapy, hospitalisation or integrated approaches) to treat these comorbid psychiatric conditions. Results from current studies present a complex aspect of the effect of the comorbidity of these disorders on treatment outcomes. The impact of AD diagnosis on SUD outcomes is complex that some studies suggest no significant impact on treatment outcomes and some suggest worse outcomes. But all of the study results agree on no clear negative effects of pharmacotherapy on anxiety outcomes. Also SUD treatment does not directly impact anxiety symptoms but in early stages of abstinence, following detoxification, anxiety symptoms significantly decrease.

Very few studies examined the safety and efficacy of pharmacotherapy for comorbid AD-SUD diagnosis. Buspirone was used in a group of opiate-dependent patients under methadone maintenance treatment, and authors did not observe any significant improvement in anxiety or substance use outcomes [53]. A series of case reports about imipramine use for the treatment of "phobic anxiety" SUD comorbidity postulated minor short-time improvements in both AD and SUD outcomes and patients who maintained the imipramine treatment for a long time had lower relapse to substance use rates [54]. On the other hand, later studies about imipramine use for cocaine and opioid dependence did not find significant benefits for the SUD-AD comorbidity, also when tolerability of selective serotonin reuptake inhibitors (SSRIs) and selective noradrenaline reuptake inhibitors (SNRIs) taken into account, they are recommended as the first-line treatment approach [55, 56].

When compared to antidepressants, benzodiazepines have the benefits of a more rapid anxiolytic effect, but their use in the treatment of SUD-AD has been controversial, due to concerns about abuse potential. In parallel with these concerns, there is a few literature examining the efficacy of safety of benzodiazepines in the treatment of AD-SUD [55]. Also, there are data in the literature suggesting that people with a personal and family history of SUDs may be more susceptible to addictive potential of benzodiazepines and thus might carry greater risk of medication abuse [56, 57]. Additionally, benzodiazepine misuse is more common in patients with SUD, especially among those with a diagnose of more sever SUD, polysubstance users and greater psychiatric comorbidity [58]. Despite all, a case series about using benzodiazepines for AD-SUD suggested that benzodiazepines were able to sustain abstinence when patients were selected carefully and monitored closely [59].

Until well-controlled studies are conducted, safety and efficacy of benzodiazepines for this population will remain unknown. Therefore, alternative therapies with better safety and efficacy profiles are recommended as the first-line therapeutic approach. Considering benzodiazepines is recommended only after these treatment options are ineffective [56, 59].

Studies have also evaluated the effect of some pharmacotherapeutic agents for comorbid AD-SUD. Topiramate, modulating principally gamma-amino butyric acid (GABA) and glutamate-mediated neurons, has been presumed as a treatment for cocaine dependence [60]. Also, there are studies suggesting topiramate as an effective treatment option in certain ADs such as SAD [61], OCD [62] and PTSD [63]. Also, studies about another GABA-ergic agent tiagabine, which has been shown to be effective in the treatment of some ADs, have conflicting results about the treatment of SUDs [64].

After reviewing general treatment considerations, authors would like to present acknowledgement about each anxiety disorder in the light of the current literature.

7. Post traumatic stress disorder

Epidemiologic studies indicate high comorbidity rates for PTSD-SUDs. The NESARC study found lifetime prevalence as 6.4 and 46.4% of people with PTSD met criteria for any SUD and 22.3% of them met criteria for substance dependence [65]. Based on the data from the Australian National Survey of Mental Health and Well-Being survey, 34.4% of individuals diagnosed with PTSD had at least one SUD, generally more than one SUD [66]. In clinical settings, when compared to patients without PTSD, patients with PTSD found to be 14 times more likely met criteria for SUD [67]. Rates of lifetime PTSD in treatment-seeking SUD populations range from 30 to 60% [68]. Studies about PTSD-SUD comorbidity indicate earlier onset of substance use, more polysubstance use, more cognitive distortions, more comorbid mental disorders, more frequent self-destructive behaviour and increased vulnerability to revictimisation [69].

Studies trying to define neurobiological mechanisms underlying this comorbidity focused on hypothalamic-pituitary-adrenal (HPA) axis and noradrenergic system. In the cerebrospinal fluid of individuals with PTSD, levels of corticotropin releasing factor (CRF) and norepinephrine were detected higher than normal levels. Considering the role of CRF and norepinephrine as a mediator of the relationship between stress and substance seeking behaviour, future research about these might explain "self-medication of PTSD symptoms using substances" theory in some individuals [70].

Although current knowledge suggests concurrent and integrated treatment strategies for patients with comorbid PTSD-SUD, studies about medications in the treatment of this comorbidity are lacking. In a prospective observational study, opioid replacement therapy was found to be effective at reducing substance use in patients with opioid dependence and PTSD. However, patients with comorbidity received higher doses of opioid medication and attended more psychosocial treatment sessions when compared to SUD-only group [71].

N-acetylcysteine (NAC), a derivative of the dietary amino acid cysteine, is used as a mucolytic agent for pulmonary diseases and used for acetaminophen toxicity. Providing an increase in the production of glutathione, it is supposed to restore substance-related glutamatergic dysregulation, and a study has shown its modest efficacy in reducing cocaine use and craving [72]. There are randomised controlled trials exploring its efficacy versus placebo in the treatment of SUD-PTSD comorbidity.

With the recent studies, the neuropeptide oxytocin has become a favourable treatment option in many psychiatric disorders. Because oxytocin is supposed to have anxiolytic and fear-modulating effects, it becomes a promising treatment option to augment exposure-based therapies for PTSD-SUD comorbidity [73]. Furthermore, with regard to neuroimaging studies, it is postulated that oxytocin might mitigate the dysregulation of corticolimbic brain circuitry, a neurobiological mechanism underlying SUD-AD comorbidity [74]. Current literature indicates that combining oxytocin treatment with psychosocial interventions may improve the treatment outcomes of this comorbidity.

Psychosocial treatment approaches can basically divide into non-exposure-based and exposure-based treatments. Seeking Safety (SS) aims to educate patients about decreasing risky behaviours, coping with substance triggers, developing self-control skills and enhancing communication skills to build a supportive environment. Transcend is an eclectic, 12-week partial hospitalisation group program and consists of 12-step treatment programs, cognitive behavioural therapy (CBT), psychodynamic and constructivist interventions. CBT for PTSD has an 8–12 session protocol that focuses on breath training, psycho-education and coping skills (e.g. cognitive restructuring and relapse prevention). Among these treatment approaches, researchers have been studied SS most thoroughly. But, there is no evidence supporting any significant positive effects of non-exposure-based treatments in the treatment of PTSD-SUD comorbidity [75].

Exposure-based psychosocial treatment approaches include prolonged exposure and concurrent treatment of PTSD and cocaine dependence. Prolonged exposure (PE) includes psycho-education, breathing training, *in vivo* exposure and imaginal exposure. Concurrent treatment of PTSD and cocaine dependence (CTPCD) is a 16-session treatment program. In first 5 sessions patients complete coping skills training, learn PTSD-focused psycho-education and PE treatment rationale. After one session focusing on imaginal and *in vivo* exposure, coping skills treatment is continued throughout the treatment protocol. According to a systematic review by Dam et al., exposure-based psychosocial treatment interventions might be more promising in concurrent treatment for PTSD-SUD comorbidity [75].

8. Panic disorder

Lifetime prevalence of panic disorder was 4.7%, and 12-month prevalence of panic disorder was 2.7% in United States [76]; people with lifetime SUD diagnosis were found to be 1.3 times more likely to meet criteria for panic disorder [3]. But there is no study or case report in the literature about the treatment of PD-SUD comorbidity.

9. Social anxiety disorder

According to NESARC data, prevalence rate of SAD-SUD comorbidity is 2.4% in general population [77]. Book et al. found that the presence of SAD had a significant negative effect on treatment motivation [78]. In their later study, they also postulated that there were no differences in the anxiety levels of people seeking treatment with comorbid SAD-SUD and SAD-only, but they also suggested the clinicians to be aware of the difficulties in engaging SAD-SUD patients in social communications [79]. There is only one case report about pharmacotherapeutic approach of SAD-SUD, and gabapentin was found to be effective in a patient with SAD-SUD comorbidity [80].

10. Obsessive-compulsive disorder

There is not much study in the literature about OCD-SUD comorbidity. In their study, False-Stewart and Schafer found that symptoms of OCD were subtle for the patient, and the patients might have the tendency to hide their symptoms. They also postulated integrated OCD/SUD physchotherapeutic interventions as an effective treatment approach for the comorbidity [48].

11. Generalised anxiety disorder

People with lifetime SUD diagnosis were found to be 1.3 times more likely to meet criteria for GAD [3], and the presence of GAD has a negative impact on treatment outcomes. Because GAD symptoms may be similar to the symptoms of intoxication or withdrawal, it is recommended to delay assessment of GAD until this period ends. No clinical trials of pharmacotherapuetic interventions on GAD-SUD have been conducted.

There are no data available in the literature about the comorbidity of novel psychoactive substances (synthetic cannabinoids, designer drugs, etc.) use and any anxiety disorder. Acute or chronic use of some of these substances is associated with anxiety symptoms and some people with anxiety disorders use these substances in order to relieve their complaints. Basically, this mutual relationship might be perceived as common etiologic origins but well-designed studies about this comorbidity should be conducted in the recent future.

12. Conclusion

In the last few decades, the close relationship between SAD and SUD has been well-documented. After then, research has focused on diagnosing comorbidity and tailoring treatment for the index patients. Today, most of the patients have the opportunity to receive treatment from both mental and addiction services. Sometimes patients are directed from one clinic to the other based on the current presenting symptoms of referral.

Although patients with SUD-AD comorbidity face with increased distress, social/occupational impairment and complicated clinical course, there is limited knowledge about effective treatment approaches. Future research should focus on developing integrated psychosocial treatments and pharmacological interventions. Discovering underlying neurobiologic mechanisms may help clinicians better understand prognostic and diagnostic parameters of this pathophysiology and treatment outcomes.

Author details

Onat Yilmaz[1,*] and Nesrin Dilbaz[2]

*Address all correspondence to: hilmiyenesrin.dilbaz@uskudar.edu.tr

1 Department of Psychiatry, Uskudar University, Istanbul, Turkey

2 Okmeydani Research and Education Hospital, Istanbul, Turkey

References

[1] Grant BF, Stinson FS, Dawson DA, Chou PS, Dufour MC, Compton W, Pickering RP, Kaplan K. Prevalence and co-occurrence of substance use disorders and independent mood and anxiety disorders: results from the National Epidemiologic Survey on Alcohol and Related Conditions. Alcohol Res Health. 2006; 29(2): 107–120.

[2] Grant BF, Goldstein RB, Chou SP, Huang B, Stinson FS, Dawson DA, Saha TD, Smith SM, Pulay AJ, Pickering RP, Ruan WJ, Compton WM. Sociodemographic and psychopathologic predictors of first incidence of DSM-IV substance use, mood and anxiety disorders: results from the Wave 2 National Epidemiologic Survey on Alcohol and Related Conditions. Mol Psychiatry. 2009; 14: 1051–1066.

[3] Grant BF, Saha T, Ruan J, Goldstein RB, Chou P, Jung J, Zhang H, Smith SM, Pickering RP, Huang B, Hasin DS. Epidemiology of *DSM-5* drug use disorder results from the

National Epidemiologic Survey on Alcohol and Related Conditions – III. JAMA Psychiatry. 2016; 73(1): 39–47.

[4] Conway KP, Compton W, Stinson FS, Grant BF. Lifetime comorbidity of DSM-IV mood and anxiety disorders and specific drug use disorders: results from the National Epidemiological Survey on Alcohol and Related Conditions. J Clin Psychiatry. 2006; 67(2): 247–257.

[5] Compton WM, Thomas YF, Stinson FS, Grant BF. Prevalence, correlates, disability, and comorbidity of DSM-IV drug abuse and dependence in the United States: results from the National Epidemiologic Survey on Alcohol and Related Conditions. Arch Gen Psychiatry. 2007; 64(5): 566–576.

[6] Mericle AA, Park VMT, Holck P, Arria AM. Prevalence, patterns, and correlates of co-occurring substance use and mental disorders in the United States: variations by race/ethnicity. Compr Psychiatry. 2012; 53: 657–665.

[7] McEvoy PM, Grove R, Slade T. Epidemiology of anxiety disorders in the Australian general population: findings of the 2007 Australian National Survey of Mental Health and Wellbeing. Aust N Z J Psychiatry. 2011; 45: 957–967.

[8] Merikangas KR, Mehta RL, Molnar BE, Walters EE, Swendsen JD, Aguilar-Gaziola S, Bijl R, Borges G, Caraveo-Anduaga JJ, DeWit DJ, Kolody B, Vega WA, Wittchen HU, Kessler RC. Comorbidity of substance use disorders with mood and anxiety disorders: results of the International Consortium in Psychiatric Epidemiology. Addict Behav. 1998; 23: 893–907.

[9] Leray E, Camara A, Drapier D, Riou F, Bougeant N, Pelissolo A, Lloyd KR, Bellamy V, Roelandt JL, Millet B. Prevalence, characteristics and comorbidities of anxiety disorders in France: results from the "Mental Health in General Population" survey (MHGP). Eur Psychiatry. 2011; 26: 339–345.

[10] Toftdahl NG, Nordentoft M, Hjorthøj C. Prevalence of substance use disorders in psychiatric patients: a nationwide Danish population-based study. Soc Psychiatry Psychiatr Epidemiol. 2016; 51: 129–140.

[11] Lai HMX, Cleary M, Sitharthan T, Hunt GE. Prevalence of comorbid substance use, anxiety and mood disorders in epidemiological surveys, 1990–2014: a systematic review and meta-analysis. Drug Alcohol Depend. 2015; 154: 1–13.

[12] Franken IHA, Hendriks VM. Screening and diagnosis of anxiety and mood disorders in substance abuse patients. Am J Addict. 2001; 10(1): 30–39.

[13] Bakken K, Landheim AS, Vaglum P. Substance-dependent patients with and without social anxiety disorder: occurrence and clinical differences. A study of a consecutive sample of alcohol-dependent and poly-substance-dependent patients treated in two counties in Norway. Drug Alcohol Depend. 2005; 80 (3): 321–328.

[14] Bakken K, Landheim AS, Vaglum P. Axis I and II disorders as long-term predictors of mental distress: a six-year prospective follow-up of substance- dependent patients. BMC Psychiatry. 2007; 7: 29.

[15] Sbrana A, Bizzarri JV, Rucci P, Gonnelli C, Doria MR, Spagnolli S, Ravani L, Raimondi F, Dell'Osso L, Cassano GB. The spectrum of substance use in mood and anxiety disorders. Compr Psychiatry. 2005; 46(1): 6–13.

[16] Rosen D, Smith ML, Reynolds CF. The prevalence of mental and physical health disorders among older methadone patients. Am J Geriatr Psychiatry. 2008; 16(6): 488–497.

[17] Smith JP, Book SW. Comorbidity of generalised anxiety disorder and alcohol use disorders among individuals seeking outpatient substance abuse treatment. Addict Behav. 2010; 35(1): 42–45.

[18] Goldner EM, Lusted A, Roerecke A, Rehm J, Fischer B. Prevalence of Axis-1 psychiatric (with focus on depression and anxiety) disorder and symptomatology among nonmedical prescription opioid users in substance use treatment: systematic review and meta-analyses. Addict Behav. 2014; 39: 520–531.

[19] Greeley J, Oei T. Alcohol and tension reduction. Psychol Theor DRrink Alcohol.Leonard KE and Blane HT (Eds) Psychological Theories of Drinking and Alcoholism. Guilford Press, NY, 1999; 14–53.

[20] Khantzian EJ. The self-medication hypothesis of addictive disorders: focus on heroin and cocaine dependence. Am J Psychiatry. 1985; 142(11): 1259–1264.

[21] Sher KJ, Levenson RW. Risk for alcoholism and individual differences in the stress-response-dampening effect of alcohol. J Abnorm Psychol. 1982; 91(5): 350–367.

[22] Breslau N, Davis GC, Schultz LR. Posttraumatic stress disorder and the incidence of nicotine, alcohol, and other drug disorders in persons who have experienced trauma. Arch Gen Psychiatry. 2003; 60: 289–294.

[23] Robinson J, Sareen J, Cox BJ, Bolton JM. Role of self-medication in the development of comorbid anxiety and substance use disorders. Arch Gen Psychiatry. 2011; 68(8): 800–807.

[24] Swendsen J, Conway KP. Mental disorders as risk factors for substance use, abuse, and dependence: results from the 10-year follow-up of the National Comorbidity Survey. Addiction. 2010; 105: 1117–1128.

[25] Buckner JD, Schmidt NB, Lang AR, Small JW, Schlauch RC, Lewinsohn PM. Specificity of social anxiety disorder as a risk factor for alcohol and cannabis dependence. J Psychiatr Res. 2008; 42(3): 230–239.

[26] Zvolensky MJ, Cougle JR, Johnson KA, Bonn-Miller MO, Bernstein A. Marijuana use and panic psychopathology among a representative sample of adults. Exp Clin Psychopharmacol. 2010; 18(2): 129–134.

[27] Brady KT, Dansky BS, Sonne SC, Saladin ME. Posttraumatic stress disorder and cocaine dependence. Order of onset. Am J Addict. 1998; 7(2): 128–135.

[28] Stewart SH, Conrod PJ. Anxiety disorder and substance use disorder co-morbidity: common themes and future directions. S.U. Disorders. New York, NY: Springer, 2008: 239–257.

[29] Hodgson K, Almasy L, Knowles EEM, Kent JW, Curran JE, Dyer TD, Göring HHH, Olvera RL, Fox PT, Pearlson GD, Krystal JH, Duggirala R, Blangero J, Glahn DC. Genome-wide significant loci for addiction and anxiety. Eur Psychiatry. 2016; 36: 47–54.

[30] Zanarini MC, Skodol AE, Bender D, Dolan R, Sanislow C, Schaefer E, Morey LC, Grilo CM, Shea MT, McGlashan TH, Gunderson JG. The collaborative longitudinal personality disorders study: reliability of axis I and II diagnoses. J Pers Disord. 2000; 14(4): 291–299.

[31] Kranzler HR, Kadden RM, Babor TF, Tennen H, Rounsaville BJ. Validity of the SCID in substance abuse patients. Addiction. 1996; 91(6): 859–868.

[32] Kranzler HR, Kadden RM, Burleson JA, Babor TF, Apter A, Rounsaville BJ. Validity of psychiatric diagnosis in patients with substance use disorders: is the interview more important than the interviewer? Compr Psychiatry. 1995; 36(4): 278–288.

[33] Hasin D, Grant BF. Diagnosing depressive disorders in patients with alcohol and drug problems: a comparison of the SAD-L and the DIS. J Psychiatr Res. 1987; 21: 301–311.

[34] Hasin D, Samet S, Nunes E, Meydan J, Matseoane K, Waxman R. Diagnosis of comorbid psychiatric disorders in substance users assessed with the psychiatric research interview for substance and mental disorders for DSM-IV. Am J Psychiatry. 2006; 163: 689–696.

[35] Hasin D, Greenstein E, Aivadyan C, Stohl M, Aharonovich E, Saha T, Goldstein R, Nunes EV, Jung J, Grant BF. The Alcohol Use Disorder and Associated Disabilities Interview Schedule-5 (AUDADIS-5): procedural validity of substance use disorders modules through clinical re-appraisal in a general population sample. Drug Alcohol Depend. 2015; 148: 40–46.

[36] Haro JM, Arbabzadeh-Bouchez S, Brugha TS, de Girolamo G, Guyer ME, Jin R, Lepine JP, Mazzi F, Reneses B, Vilagut G, Sampson NA, Kessler RC. Concordance of the Composite International Diagnostic Interview Version 3.0 (CIDI 3.0) with standardized clinical assessments in the WHO World Mental Health surveys. Int J Methods Psychiatr Res. 2006; 15(4): 167–180.

[37] Castle DJ. Anxiety and substance use: layers of complexity. Expert Rev Neurotherapeutics. 2008; 8(3): 493–501.

[38] Ouimette PC, Gima K, Moos RH, Finney JW. A comparative evaluation of substance abuse treatment IV. The effect of comorbid psychiatric diagnoses on amount of treatment, continuing care, and 1-year outcomes. Alcohol Clin Exp Res. 1999; 23(3): 552–557.

[39] Ford JD, Hawke J, Alessi S, Ledgerwood D, Petry N. Psychological trauma and PTSD symptoms as predictors of substance dependence treatment outcomes. Behav Res Ther. 2007; 45(10): 2417–2431.

[40] Mills KL, Teesson M, Ross J, Darke S. The impact of post-traumatic stress disorder on treatment outcomes for heroin dependence. Addiction. 2007; 102(3): 447–454.

[41] Tomasson K, Vaglum P. The 2-year course following detoxification treatment of substance abuse: the possible influence of psychiatric comorbidity. Eur Arch Psychiatry Clin Neurosci. 1997; 247(6): 320–327.

[42] Brown PJ, Read JP, Kahler CW. Comorbid posttraumatic stress disorder and substance use disorders: treatment outcomes and the role of coping. In Ouimette, Paige, Brown, Pamela J. (Eds). (2003). Trauma and substance abuse: Causes, consequences, and treatment of comorbid disorders , American Psychological Association, Washington, DC, 1993: 171–188.

[43] Brown PJ. Outcome in female patients with both substance use and post-traumatic stress disorders. Alcohol Treat Q. 2000; 18(3): 127–135.

[44] Ouimette PC, Humphreys K, Moos RH, Finney JW, Cronkite R, Federman B. Self-help participation and functioning among substance use disorder patients with PTSD. J Subst Abuse Treat. 2001; 20: 25–32.

[45] Ouimette PC, Moos RH, Finney JW. PTSD treatment and 5-year remission among patients with substance use and posttraumatic stress disorders. J Consult Clin Psychol. 2003; 71(2): 410–414.

[46] Ouimette PC, Finney JW, Moos RH. Two-year posttreatment course and coping of patients with substance abuse and posttraumatic stress disorder. Psychol Addict Behav. 1999; 13: 105–114.

[47] Regier DA, Farmer ME, Rae DS, Locke BZ, Keith SJ, Judd LL, et al. Comorbidity of mental disorders with alcohol and other drug abuse: results from the Epidemiologic Catchment Area (ECA) Study. JAMA. 1990; 264: 2511–2518.

[48] Fals-Stewart W, Schafer J. The treatment of substance abusers diagnosed with obsessive-compulsive disorder: an outcome study. J Subst Abuse Treat. 1992; 9: 365–370.

[49] Bruce S, Yonkers K, Otto M, Eisen J, Weisberg R, Pagano M, et al. Influence of psychiatric comorbidity on recovery and recurrence in generalized anxiety disorder, social phobia, and panic disorder: a 12-year prospective study. Am J Psychiatry. 2005; 162: 1179–1187.

[50] Fatseas M, Denis C, Lavie E, Auriacombe M. Relationship between anxiety disorders and opiate dependence—a systematic review of the literature: Implications for diagnosis and treatment. J Subst Abuse Treat. 2010; 38: 220–230.

[51] Substance Abuse and Mental Health Services Administration, Office of Applied Studies. The NSDUH Report: adolescent mental health: service settings and reasons for receiving care (February 19, 2009). Rockville, MD, 2009. Available at: http://www.oas.samhsa.gov/2k9/youthMHcare/youthMHcare.htm.

[52] Sterling S, Weisner C, Hinman A, Parthasathy S. Access to treatment for adolescents with substance use and co-occurring disorders: challenges and opportunities. J Am Acad Child Adolesc Psychiatry. 2010; 49(7): 637–646.

[53] McRae AL, Sonne SC, Brady KT, Durkalski V, Palesch Y. A randomized, placebo-controlled trial of buspirone for the treatment of anxiety in opioid-dependent individuals. Am J Addict. 2004; 13: 53–63.

[54] Quitkin FM, Rifkin A, Kaplan J, Klein DF. Phobic anxiety syndrome complicated by drug dependence and addiction. A treatable form of drug abuse. Arch Gen Psychiatry. 1972; 27: 159–162.

[55] Nunes EV, McGrath PJ, Quitkin FM, et al. Imipramine treatment of cocaine abuse: possible boundaries of efficacy. Drug Alcohol Depend. 1995; 39: 185–195.

[56] Ciraulo DA, Nace EP. Benzodiazepine treatment of anxiety or insomnia in substance abuse patients. Am J Addict. 2000; 9: 276–9; discussion 280–284.

[57] Ciraulo DA, Sarid-Segal O, Knapp C, Ciraulo AM, Greenblatt DJ, Shader RI. Liability to alprazolam abuse in daughters of alcoholics. Am J Psychiatry. 1996; 153: 956–968.

[58] Busto UE, Romach MK, Sellers EM. Multiple drug use and psychiatric comorbidity in patients admitted to the hospital with severe benzodiazepine dependence. J Clin Psychopharmacol. 1996; 16: 51–57.

[59] Adinoff B. Long-term therapy with benzodiazepines despite alcohol dependence disorder. Am J Addict. 1992; 1: 288–293.

[60] Johnson BA. Recent advances in the development of treatments for alcohol and cocaine dependence: focus on topiramate and other modulators of GABA or glutamate function. CNS Drugs. 2005; 19(10): 873–896.

[61] Tucker P, Trautman RP, Wyatt DB, et al. Efficacy and safety of topiramate monotherapy in civilian posttraumatic stress disorder: a randomized, double-blind, placebo-controlled study. J Clin Psychiatry. 2007; 68(2): 201–206.

[62] Van Ameringen M, Mancini C, Pipe B, et al. An open trial of topiramate in the treatment of generalized social phobia. J Clin Psychiatry. 2004; 65(12): 1674–1678.

[63] Rubio G, Jimenez-Arriero MA, Martinez-Gras I, et al. The effects of topiramate adjunctive treatment added to antidepressants in patients with resistant obsessive-compulsive disorder. J Clin Psychopharmacol. 2006; 26(3): 341–344.

[64] Smith JP, Book SW. Anxiety and substance use disorders: a review. Psychiatr Times. 2008; 25(10): 19–23.

[65] Pietrzak RH, Goldstein RB, Southwick SM, Grant BF. Prevalence and axis I comorbidity of full and partial posttraumatic stress disorder in the United States: results from Wave 2 of the National Epidemiologic Survey on Alcohol and Related Conditions. J Anxiety Disord. 2011; 25(3): 456–465.

[66] Mills KL, Teesson M, Ross J, Peters L. Trauma, PTSD, and substance use disorders: findings from the Australian National Survey of Mental Health and Well-Being. Am J Psychiatry. 2006; 163(4): 652–658.

[67] Ford JD, Russo EM, Mallon SD. Integrating treatment of posttraumatic stress disorder and substance use disorder. J Couns Dev. 2007; 85: 475–489.

[68] Brady KT, Haynes LF, Hartwell KJ, Killeen TK. Substance use disorders and anxiety: a treatment challenge for social workers. Soc Work Public Health. 2013; 28: 407–423.

[69] Schafer I, Najavits LM. Clinical challenges in the treatment of patients with posttraumatic stress disorder and substance abuse. Curr Opin Psychiatry. 2007; (20): 614–618.

[70] Flanagan JC, Korte KJ, Killeen KT, Back SE. Concurrent treatment of substance use and PTSD. Curr Psychiatry Rep. 2016; 18: 70.

[71] Trafton JA, Minkel J, Humphreys K. Opioid substitution treatment reduces substance use equivalently in patients with and without posttraumatic stress disorder. J Stud Alcohol. 2006; 67: 228–235.

[72] Brown RM, Kupchik YM, Kalivas PW. The story of glutamate in drug addiction and of N-acetylcysteine as a potential pharmacotherapy. JAMA Psychiatr. 2013; 70(9): 895–897.

[73] Eckstein M, Becker B, Scheele D, Scholz C, Preckel K, Schlaepfer TE, et al. Oxytocin facilitates the extinction of conditioned fear in humans. Biol Psychiatry. 2014; 78(3): 194–202.

[74] Dodhia S, Hosanagar A, Fitzgerald DA, Labuschagne I, Wood AG, Nathan PJ, et al. Modulation of resting-state amygdala-frontal functional connectivity by oxytocin in generalized social anxiety disorder. Neuropsychopharmacology. 2014; 39(9): 2061–2069.

[75] Dam D, Vedel E, Ehring T, Emmelkamp PMG. Psychological treatments for concurrent posttraumatic stress disorder and substance use disorder: a systematic review. Clin Psychol Rev. 2012; 32: 202–214.

[76] Kessler RC, Chiu WT, Demler O, Merikangas KR, Walters EE. Prevalence, severity, and comorbidity of 12-month DSM-IV disorders in the National Comorbidity Survey Replication. Arch Gen Psychiatry. 2005; 62(6): 617.

[77] Schneier FR, Foose TE, Hasin DS, Heimberg RG, Liu SM, Grant BF, et al. Social anxiety disorder and alcohol use disorder co-morbidity in the National Epidemiologic Survey on Alcohol and Related Conditions. Psychol Med. 2010; 40(6): 977–988.

[78] Book SW, Thomas SE, Dempsey JP, Randall PK, Randall CL. Social anxiety impacts willingness to participate in addiction treatment. Addict Behav. 2009; 34(5): 474–476.

[79] Book SW, Thomas SE, Smith JP, Miller PM. Severity of anxiety in mental health versus addiction treatment settings when social anxiety and substance abuse are comorbid. Addict Behav. 2012; 37(10): 1158–1161.

[80] Verduin ML, McKay S, Brady KT. Gabapentin in comorbid anxiety and substance use. Am J Addict. 2007; 16(2): 142–143.

Chapter 6

Parental Involvement in Remotely Delivered CBT Interventions for Anxiety Problems in Children and Adolescents: A Systematic Review

Costina-Ruxandra Păsărelu and Anca Dobrean

Additional information is available at the end of the chapter

http://dx.doi.org/10.5772/65120

Abstract

Introduction: Remotely delivered interventions for childhood anxiety disorders (e.g., delivered via telephone, Internet, computer, serious games, or apps) are efficient in mental health problems, surpassing concerns in the dissemination of evidence-based treatment. The present study aims to conduct a systematic review of parental involvement in remote cognitive-behavioral therapy (CBT) interventions for child anxiety disorders. The main objectives are (1) to present the state of the art of existent knowledge on parental involvement in remotely delivered CBT interventions for anxiety disorders in children and adolescents and (2) to propose a conceptual model which could be considered in designing effective remotely delivered interventions for anxious youth.

Research methods: Two independent raters conducted comprehensive systematic searches of electronic databases (PsychInfo, Cochrane, PubMed, Scopus, Web of Science) up to April 10, 2016. We have included eight remote CBT interventions and presented their characteristics.

Conclusion: Parental involvement in remotely delivered CBT interventions for child anxiety treatment is important to consider; however, more randomized controlled trials need to be conducted. Tailoring such interventions according to child-parent factors, child-parent relationship quality, and treatment preferences has been suggested. Finally, a new conceptual model has been proposed, which could guide future research.

Keywords: anxiety, child, remote, CBT

1. Introduction

Anxiety disorders are the most widespread mental health problems in children and adolescents, affecting approximately 117 million of youths, with a worldwide prevalence of any anxiety disorder of 6.5% [1]. Anxiety disorders in youths are associated with a high burden of disease, being the second leading cause of disability-adjusted life years and years lived with disability rates, after major depressive disorder for girls, and the third leading cause of disability for boys, after major depressive disorder and conduct disorder [2]. Childhood anxiety disorders often persist in adulthood [3], becoming a chronic condition, so that anxious children turn into anxious adults with impaired functioning in several psychosocial domains [4].

2. Evidence-based treatments for child and adolescent anxiety

2.1. Existent treatments

At present, according to mental health guidelines [5], the first-line treatment delivered in youth anxiety disorders is cognitive-behavioral therapy (CBT), which is an efficient intervention for youths [6]. However, a high number of youths are unresponsive to treatment (40%) [6]. Even though evidence-based treatments exist, there are many problems in reaching these populations: trained therapists, costs, rural areas, or stigma associated with going to a mental health clinic. Due to current problems in disseminating evidence-based treatments, remotely delivered interventions have emerged and proved to be efficient treatments for multiple conditions, both in adults [7] and in youth populations [8].

2.1.1. Remote treatment for child anxiety

There is a wide variety of remotely delivered treatments for child and adolescents with anxiety disorders [9]. Remote interventions can be delivered over Internet, computer, telephone, SMS, mobile apps, videoconferences, serious games, and virtual reality devices. In the last years, there has been a surge in such interventions due to people's greater access to technology.

2.1.1.1. Internet-delivered interventions

Internet-delivered CBT interventions (iCBT) are those treatments in which the content of the treatment is presented online (for a synthesis of iCBT interventions see [10]). There is evidence that iCBT is an efficient intervention for anxiety disorders in youths of all ages, from preschoolers to adolescents. Namely, in a pilot study conducted on children with specific phobias, iCBT was efficient in reducing the main symptoms [11]. The same protocol was tested in a randomized controlled trial (RCT) for children with anxiety disorders and seemed to be an efficient treatment [12]. Also, iCBT treatments considerably reduced social phobia in a sample group of adolescents [13]. A feasibility study showed that iCBT delivered in a tailored format was an efficient treatment in a sample group of adolescents with mixed anxiety disorders [14]. These results were also sustained by an RCT conducted on children with mixed anxiety disorders

[15]. Other studies bring evidence that iCBT treatments for children can be as effective as golden standards, i.e., face-to-face CBT delivered in a clinic [16]. Online-delivered CBT is also efficient in anxious preschool children, albeit it is delivered mostly as a parenting intervention [17, 18].

2.1.1.2. Computerized interventions

Computerized CBT interventions (cCBT) are often delivered over CD-ROMs. Several trials have shown that cCBT are efficient interventions with comparable results to face-to-face treatments [19]. In an adolescent sample of anxious youths, cCBT delivered over 12 weeks was more efficient than waiting control [20]. When compared with treatment as usual (TAU), cCBT interventions are associated with better outcomes than TAU [21].

2.1.1.3. Intervention delivered via mobile phones, smartphones

The usability of two mobile CBT interventions (mCBT) has been investigated so far, with future RCTs still required. In a pilot study, the feasibility of SmartCAT, a smartphone-enhanced child anxiety program used as a complementary tool to a brief CBT intervention, was investigated [22]. SmartCAT consists of a smartphone app, a therapist portal, and a two-way communication connection between them. Data on the usability of this intervention show that the app was frequently used during the first week, while the level of usage decreased during treatment, the compliance rate was high, and it was considered easy to use. Parents reported a high level of satisfaction with the intervention. Mayo Clinic Anxiety Coach [23] is a self-help application on iOS, consisting of three modules: Self-Evaluation, Psychoeducation, and Exposure. It was downloaded by approximately 169 children, aged between 5 and 17, and used multiple times (4–20 times). The first module (self-assessment) was used more than the other two. The Pesky gNATs mobile app [24] is a free access tool, available for iPhone and Android phones and tablets, which can be used by children in order to generalize contents learnt in therapy sessions. This app has not yet been tested in RCT, or in feasibility trials.

2.1.1.4. Serious games

In a two-armed RCT, Dojo, a videogame targeting anxiety symptoms in teenagers (aged 11–15) was compared with a control game, Rayman [25]. No differences were found between the two conditions. Treasure Hunt is a CBT game delivered over six levels, which was liked by children and rated as helpful by therapists [26]; however, no RCT has had its efficacy tested compared to control conditions. MindSpace is a CBT game for children aged 7–12 diagnosed with anxiety disorders [27]. Pesky gNATs [24] is a computer game and an online app designed to be used as a complementary tool in therapy. There are seven levels in which the anxious child learns CBT concepts. However, the efficacy of Pesky gNATs has not been investigated so far in experimental or controlled settings.

2.1.1.5. Teleconferencing/chat

Telepsychiatry refers to psychiatric applications using videoconferencing communication. The TeleLink Mental Health Program (TeleLink) is a feasible manner in which children with mental health problems from remote areas can be treated [28]; supervision-based telepsychiatry for therapists can also be delivered [29].

2.1.1.6. SMS/e-mail

'Reach Out, Rise Up' is a text message intervention that significantly reduced anxiety and depression scores in a sample group of youths [30]. The intervention consisted of sending three text messages each week, for a period of 10 weeks: one message regarding psychoeducation content, one challenge related to the content learnt, and an inspirational message.

2.1.1.7. Interventions delivered via virtual environments, virtual reality

Virtual reality (VR) represents a set of technologies that allow an individual to interact in real time with a 3D virtual environment. There are many interventions in virtual reality delivered for children and adolescents with anxiety disorders. In a feasibility study on VR therapies for social phobia in children, parents, children, and therapists rated such environments as acceptable and useful tools [31]. St-Jacques [32] compared a combined in vivo with VR exposure with exposure in vivo for children with arachnophobia, in order to investigate differences in motivation levels in children. There were no differences between the two conditions in terms of motivation; however, there were several interactions with parts of the treatment. Gutiérrez-Maldonado [33] tested VR exposure for children with school phobia in a sample group of children aged between 10 and 15, compared with no treatment. Despite the fact that school-related fears diminished considerably, general anxiety was not changed.

2.2. Parental involvement in offspring's treatment

There is mixed evidence regarding parental involvement in child CBT treatment, while there is proof that parental involvement does not add significantly to a child's therapy [6, 34], recent studies show that parental involvement in child therapy is associated with better outcomes in terms of remission rates at follow-up [35]. In a study focusing on in-session as well as out-session activities, there emerged some differences (high involvement in the former, low in the latter) that could stand for possible explanations for the mixed findings regarding added benefits of parental involvement in child anxiety treatment [36]. So far, no qualitative or quantitative approaches have been conducted in order to investigate/summarize parental involvement in remotely delivered CBT interventions with anxious children and adolescents. Such an approach is needed due to the fact that there are many remote interventions targeting mental health problems in youths which differ in several characteristics. Apart from characteristics related to the number of modules/sessions delivered, content, multimedia presentations, therapists involved (with or without clinician guidance), there are also differences with regard to parental involvement. While in several studies on remote CBT interventions, parents are mentioned only because they are the legal caregivers of the youth and they offer parental consent on their child's involvement in treatment, and are sometimes involved in the youths' mental health assessments (self-reports and clinician assessments) at the beginning and end of treatment, in other trials parents are actively involved in youth therapy. In these trials, parents read materials, help their children cope with anxiety, and have phone sessions with clinicians. However, it is unclear whether parents should be involved in youth remote CBT anxiety treatment.

2.2.1. Aim/objectives

The main objectives of the current review are: (1) to offer the state of the art of existent knowledge on parental involvement in remotely delivered CBT interventions for anxiety disorders in children and adolescents and (2) to propose a conceptual model which could be considered in designing effective remotely delivered interventions for anxious youth.

3. Method

The current systematic review has been conducted in accordance with the Preferred Reporting Items for Systematic Reviews and Meta-Analyses (PRISMA) statement [37].

3.1. Inclusion and exclusion criteria

We included only (a) randomized controlled trials with (b) remotely delivered (c) cognitive-behavioral interventions for (d) youths (children and adolescents), with (e) either a primary diagnosis and/or elevated symptoms of anxiety, which also involved (f) parents in the intervention.

We excluded trials in which the intervention did not include parental involvement in the treatment, the protocol was not CBT-oriented, or if the intervention was delivered exclusively for parents or conducted with adult participants. Also, trials in which the primary disorder was not anxiety, or targeted a primary medical condition, were excluded. Studies analyzing secondary data, systematic reviews/theoretical synthesis of literature, and feasibility studies were excluded. We selected only those studies published in English. There was no limitation regarding the year of publication.

3.2. Search strategy

Electronic databases (PsychInfo, PubMed, Scopus, Web of Science) were consulted by two independent assessors up to April 10, 2016. We combined various terms related to the means of delivery, with terms pertaining to anxiety, as well as various terms indicating CBT treatments, and terms germane to children and adolescents. Finally, all these terms were combined with those related to randomized controlled trials. The reference lists of pertinent papers and meta-analyses published on the subject were screened in order to identify potentially relevant articles. We repeated the search on July 25, 2016, to double-check and identify whether new articles had been published since the first search.

3.3. Quality assessment

We assessed the quality of the studies included using the Risk of Bias' tool, developed by Cochrane Collaboration [38]. We included the following sources of bias: random sequence generation (selection bias), allocation concealment (selection bias), blinding outcome assessments (detection bias), and the incomplete outcome data (attrition bias). We did not include blinding of participants and personnel (performance bias) due to the fact that in remotely delivered treatments, most of the time, the therapists involved cannot be blinded regarding patients' allocation.

4. Results

4.1. Literature search

The literature search resulted in 3121 records. After removing duplicates, we screened the titles and abstracts of 2340 papers. Of these, 2298 papers were rejected and we further read the full text of 42 papers. Thirty-four papers were rejected due to the following reasons: no parental involvement (n = 12), interventions delivered exclusively with parents of anxious children (n = 2), delivered only with adult participants (n = 8), targeting a primary medical condition (n = 1), not CBT (n = 1), secondary analysis (n = 1), not RCT (n = 5), systematic reviews (n = 2), primary disorder not anxiety (n = 2). Subsequently, a total of eight RCTs were included in the synthesis. **Figure 1** presents the flowchart describing the inclusion of studies.

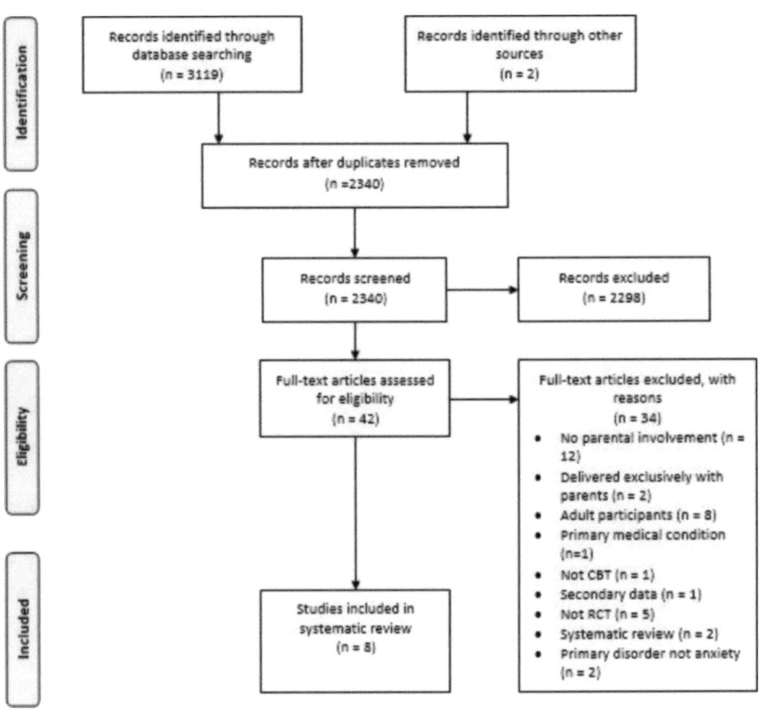

Figure 1. Flow diagram of study selection process.

Study name Country	Condition	N % female participants	M_{age} (SD) Age range	Diagnosis	Program	Adherence	Parental involvement	Parents characteristics	Therapist
Infantino et al., 2016 Australia	Audio-based vs WL	24 Children (54% girls)	7.25 (.49) 5-11	SAD (16.67%) SoP (25%) GAD (33.33%) SP (16.67%) OCD (8.03%)	Turnaround 10 lessons (20-30 min) over 5 weeks	No dropouts no missing data	-listen (together with the family or independently) -familiarize with core CBT in order to assist the child -offered two CDs (psychoeducation, praise, problem-solving strategies)	66.66% audio group parent completed university Age mothers 40.17 (1.18) fathers 43.42 (1.67) 66.66% from high income > 100.000$	No therapist contact
Khanna & Kendall, 2010 SUA	cCBT vs individual CBT, CESA	49 Children (32% girls)	10.1 7-13	GAD (57.1%) SP (16.3%) SAD (14.3%) SP (8.1%) PD (4%)	Camp Cope-A-Lot 12 sessions (35 min)	The 45 remaining participants completed all 12 sessions within 15 weeks of their first session.	-2 parent sessions conducted by the coach while youth worked independently on Levels 3 & 7	No data on parents provided	A coach involved in exposure sessions (first 6 are completed independently, next 6 completed assisted by the therapist)
March et al., 2009 Australia	iCBT vs CBT face-to-face vs CESA	73 Children (54% girls)	9.45 (1.37) 7-12	SAD (32.5%), SoP (37.5%), GAD (20%), SP (10%)	BRAVE for Children – ONLINE 10 weekly 60 min sessions (child) 6 60 min sessions (parent) 2 booster sessions	Post-test 60% parents 33.3 % child Follow-up 72.3% parents 62% child	-parent sessions focused on core CBT	22.5% mothers 20.6% fathers completed university 21.1% iCBT 31,1% WL high income >100.000 $ Age mothers 41.75 (5.34) fathers 44.78 (6.25)	Weekly contact 2 telephone contacts automated e-mails
McGrath et al., 2011 Canada	tCBT vs TAU	Anxiety sample 91 Children (64.15% girls)	8.82 (1.69) 6-12	28% tCBT 24.4% from control had comorbid disorders	The Chase Worries Away intervention 11 sessions 2 booster session	95% for session 86% for video 90% for skill implementation	-Coaches reinforced the handbook and video content with the parent and/or child, helped the family solve problems, and provided encouragement	32% parents from tCBT 24.4% from control completed university 18% >55.000$ Age 19-46	Telephone sessions from a coach (40 min)

Spence et al., 2011 Australia	iCBT vs CLIN vs WL	115 (59.13% girls)	13.98 (1.63) 12-18	GAD (48%) SoP (35%) SAD (13%) SP (4%)	BRAVE for Teenagers—ONLINE 10 weekly 60 min sessions (adolescents) 5 60 min sessions (parent) 2 booster sessions	79% parents 57 % adolescents	- parent sessions focused on core CBT + parenting strategies	46% fathers, 56% mothers completed university	15 min telephone call after session 5 to guide in exposure Monitors progress Sends e-mails after each session Automated e-mails
Storch et al, 2015 SUA	cCBT vs TAU	100 (44% girls)	9.82 (1.82) 7-13	GAD (44.9 %) PD (2%) SAD (18.4 %) SoP (26.5 %) SP (8.2 %)	Camp Cope-A-Lot 50-60 min sessions 12 weeks		- psychoeducation and parental support on sessions 3 & 7	51% parents graduated college 22.2% high income > 90.000$	Coach involved in exposure sessions
Vigerland et al, 2016 Sweden	iCBT vs WL	93 (55% girls)	10.1 (1.7) 8-12	GAD (22%) PD (5%) SAD (32%) SoP (10%) SP (31%)	11 modules delivered over 10 weeks	83% families completed the first 9 modules 63% completed at least one of the two maintenance modules	-had to complete 7 modules containing information on how to help their anxious child	Ages 31-60 43.1 (5.3) 67% had a university degree	Online contact (written messages, written feedback on worksheets) telephone calls
Wuthrich et al., 2012 Australia	cCBT vs WL	43 (62.79% girls)	15.17 (1.11) 14-17	GAD (37.5%) SoP (50%) SAD (4.2%) OCD (10.5%) SP (10.5%) ANOS (0%)	Cool Teens 12 weeks 8 modules (30 min)	98.4% telephone call compliance rate	-parental involvement decided by the adolescent - assisting parents with handouts of CBT core strategies -telephone sessions	60% high family income >80.000$	Telephone sessions with adolescent and parent

Note. cCBT =computerized cognitive-behavioral therapy; tCBT=telephone-based cognitive-behavioral therapy; iCBT= Internet-delivered cognitive-behavioral therapy; WL = waitlist; TAU = treatment as usual; GAD = generalized anxiety disorder; PD = panic disorder; SAD = separation anxiety disorder; SoP=social phobia; SP = specific phobia; OCD = obsessive compulsive disorder; ANOS = anxiety disorder without other specifications

Table 1. Characteristics of the studies included.

4.2. Study characteristics

A summary of the characteristics of the studies included (number of participants, age, type of remote intervention, parental involvement) is described in **Table 1**.

4.2.1. Types of interventions

March et al. [15] tested the efficacy of an iCBT intervention for child anxiety disorders. BRAVE for Children-ONLINE consisted of ten sessions for children delivered over 10 weeks, combined with six sessions delivered for parents. Booster sessions were also conducted at the end of the treatment in order to prevent relapses and to consolidate learnt skills. This intervention was delivered with therapist support; however, there was only minimal involvement, namely weekly online contact and two telephone sessions (description of the program and guidance on exposure). The credibility of the program was high; nevertheless, there was moderate satisfaction with the treatment, both in child and in parent ratings. Adherence, defined as the number of participants that completed the entire intervention, was between 33.3% (children) and 60% (parents), while at the 6-month follow-up this was higher, namely 62% (children) and 72.3% (parents).

Spence [16] investigated the efficacy of BRAVE for adolescents-ONLINE, compared with clinical-based treatment and wait-list. Adolescents had to complete ten sessions, while their parents completed five sessions. iCBT had similar outcomes to face-to-face treatment delivered in a clinic. Adherence was between 39% (adolescents) and 66% (parents). At the 12-month follow-up, 57% adolescents and 79% parents had completed all modules. There was moderate satisfaction with the treatment for both adolescents and parents; however, parents from the clinical condition reported a higher degree of satisfaction.

McGrath [39] compared a telephone-based CBT intervention with usual care for anxious children and found significant effects 240 days after the intervention had finished, with no effects after 120 days. The 'Chase Worries Away' intervention consisted of eleven CBT sessions plus two booster sessions in which a coach delivered the intervention by telephone. Parental involvement is not clearly described in this study, even though it is stated that telephone sessions with a coach were delivered with the family.

In a cCBT intervention, Wuthrich et al. [20] tested the Cool Teens program and found it to be an efficient intervention for adolescents with anxiety disorders. The intervention was delivered over 12 weeks and consisted of eight modules (30 min each) delivered on a CD-ROM. The intervention was delivered with clinician guidance, as there were several telephone sessions between therapist and adolescents, and therapist and parents. Significant improvements in anxiety outcomes were reported in clinician-rated instruments (41% did not meet primary anxiety diagnostic criteria and 23.5% did not meet criteria for no anxiety disorder on post-treatment, compared with no participants from control and 20% at follow-up), as well as in parent and child self-reports. Parental involvement in treatment was decided by the adolescent and consisted of offering handouts on CBT principles and telephone sessions.

Vigerland et al. [12] tested the efficacy of an iCBT treatment program compared with WL. The intervention consisted of 11 modules delivered over 10 weeks and involved both children (four modules) and parents (seven modules). The intervention was clinician-guided, with therapist contact delivered through written feedback and telephone. Satisfaction with the program was moderate in both parent and child ratings. Adherence to the program was not very high, given the mean number of modules completed by the families (9.7, SD 1.8, range 4–11), with 83% of families completing the first nine modules and 63% completing at least one of the two maintenance modules.

Infantino [40] tested the efficacy of an audio intervention delivered in a clinic. The intervention was delivered over 5 weeks and consisted of ten lessons with a 20- to 30-min duration. Satisfaction with the treatment was moderate to high for both children and their parents. In terms of adherence, there were no missing data and no dropouts. At postassessment, several children from the intervention group no longer met criteria for their primary anxiety diagnosis, compared with participants from the control group (58.3 vs 16.7%). At the 3-month follow-up, 66.67% of the children from the audio condition were free from their primary diagnosis and 41.67% were free of any anxiety diagnosis.

Khanna and Kandall [19] tested the efficacy of the Camp Cope-A-Lot program, a cCBT intervention, compared with individual face-to-face CBT and computer-assisted education support and attention condition (CESA). In terms of satisfaction with the treatment, there were no differences in child ratings for cCBT and individual CBT conditions. In parent ratings, there were no differences between the three conditions, albeit there were slight tendencies to prefer cCBT and individual CBT over CESA.

Storch et al. [21] tested the efficacy of Camp Cope-A-Lot in a sample group of anxious children, compared with treatment as usual. The intervention consisted of 12 sessions delivered weekly, with the first six sessions delivered exclusively on the computer, while the other six sessions involved clinician guidance. Therapist helped children and their parents in exposure to fearful stimuli sessions. There was high satisfaction with the treatment reported by parents and children.

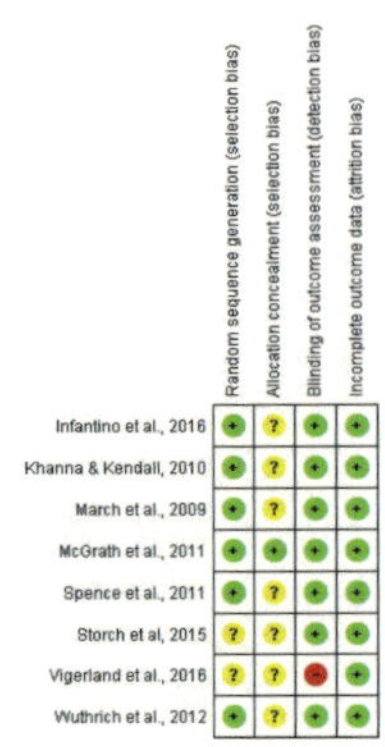

Figure 2. Risk of bias summary: review authors' judgments about each risk of bias item for each study included.

4.2.2. *Quality assessment*

Figures 2 and **3** illustrate the quality of the studies included. Most of these studies reported an adequate random sequence generation (n = 6), although in several studies the selection bias risk was unclear (n = 2). Only in one study was the allocation assignment concealed by sealed

http://dx.doi.org/10.5772/65120

double envelopes, while in the other studies allocation concealment was unclear (n = 7). In most of the studies, there were blind assessors (n = 7), while in one study there was a high risk of detection bias. There was no evidence of attrition bias, as there were either no missing data, or adequate statistical methods (intent-to-treat analysis) were used in order to control the effects of missing data. Only one of the studies included met all four of the quality criteria, while most of the trials included met two or three criteria. Two independent reviewers conducted the assessments of study quality with a high inter-rater agreement for each of the sources of risk. Disagreements were resolved via discussions until reaching consensus.

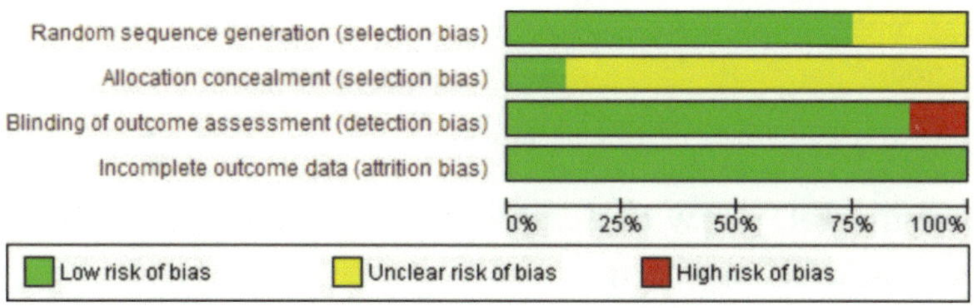

Figure 3. Risk of bias graph: review authors' judgments about each risk of bias item presented as percentages across all studies included.

4.2.3. Description of interventions (duration, modules)

We included three computerized interventions [19–21], three CBT interventions delivered over the Internet [12, 15, 16], one audio intervention [40], and one intervention delivered through telephone [39]. Youths from the studies included were aged between 5 and 17, with more female participants (54%). Children and adolescents included in the studies were diagnosed either with generalized anxiety, social anxiety, panic disorder, separation anxiety or with specific phobias, and presented mixed comorbidities (anxiety disorders, mood disorders, or externalizing disorders). Each intervention consisted of several module/lessons/sessions that ranged between 8 and 12, lasting between 20 and 60 min, and delivered between 5 and 12 weeks. Seven interventions also included a therapist/clinician, while one intervention did not include therapist guidance [40]. As we included only CBT interventions, the content was similar in each of the sessions, consisting of: psychoeducation, relaxation and breathing exercises, cognitive strategies (self-talk and cognitive restructuring), problem-solving, exposure, contingency management, and relapse prevention. The content of the interventions was presented in entertaining ways, such as text on slides, handbooks, photographs, videos, games, cartoon animation, quizzes, or audio stories of fictional characters.

All the interventions were conducted in highly developed countries: Australia (n = 4), USA (n = 2), Canada (n = 1), Sweden (n = 1). Four of the studies included compared a remotely delivered intervention with waiting list control [12, 16, 20, 40], two of the previous studies also compared remote intervention with psychoeducation [15, 19], two studies compared it with treatment as usual [21, 39], and in three studies, remotely delivered interventions were also compared with face-to-face CBT [15, 16, 19].

4.2.4. *Parental involvement*

Not all the studies reported demographic characteristics of the parents included, albeit most of the mothers and fathers included had completed higher education (university studies) and had a high annual income (expressed in several studies as more than 100,000$). Studies differed slightly in what parental involvement meant, as while several interventions are described as a combination of parent and child treatment, in which parents have to also read several modules independently, in other interventions parents are involved only in two sessions in which psychoeducation is offered. In most of the interventions, parents learn core CBT principles and strategies in order to comprehend how to help their children to acquire the skills learnt in therapy and to assist them effectively when they encounter situations in which they can become anxious.

5. Conceptual model—how to design effective remotely delivered interventions for anxious youth

Parental involvement in remotely delivered CBT interventions needs to have a sound scientific background. Despite what is currently done in trials in which parents are involved in children's treatment programs as an add-on tool, many aspects need to be considered when designing efficient remote CBT interventions for children and adolescents with parental involvement. We have not been able to find any trial in which a remote intervention delivered solely for children is compared with a control condition involving both parent and child. Also, none of the studies included assessed factors related to changes in parent variables (anxiety or comorbidities, and mechanisms of change, such as parental cognitions). Therefore, it remains unclear whether parents should be included in remote CBT interventions and, if included, what parental involvement means. Is parental involvement in child anxiety remote treatment

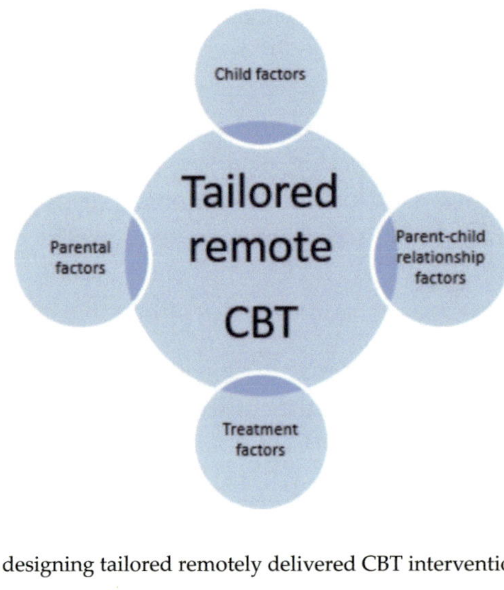

Figure 4. Conceptual model for designing tailored remotely delivered CBT interventions with parental involvement.

http://dx.doi.org/10.5772/65120

a component that needs to be present throughout the treatment or are there some important timelines in which parental involvement could be associated with greater success of the therapy (exposure sessions, generalizability of learnt skills, homework compliance)? Should parents be involved in child treatment as a function of the comorbidities their offspring present? Several factors that are proposed for consideration when involving parents in youth anxiety remote CBT are presented in **Figure 4**.

Efficient remote CBT interventions need to take into consideration an important aspect that is often not considered: tailoring or personalizing an intervention according to patients' unique characteristics. Designing efficient interventions for children needs to take into account aspects related to: parents, children, parent-child relationship, and treatment.

5.1. Child factors

Several socio-demographic characteristics (age, gender) can be considered when designing remote interventions for anxious children and adolescents. Unique mental health symptoms and disorderly clinical presentations (anxiety and comorbidities) should be tackled with a focus on tailored and transdiagnostic remote treatments. Evidence coming from literature on adults shows that transdiagnostic and tailored iCBT interventions are efficient in reducing main anxiety and depression problems, as well as in improving quality-of-life outcomes (Păsărelu et al., manuscript in preparation). Also, focusing on mechanisms of change (cognitive distortions/errors) and testing theories on which treatments are based can help us design evidence-based treatments.

5.2. Parental factors

Several factors related to parents that have been investigated so far are: parental psychopathology, several socio-demographic factors related to parents (gender, education, income, and family structure), parenting styles, and parental accommodation. Parental psychopathology is related to treatment outcomes in anxious children [41], via family functioning and caregiver strain [42]. It appears that gender and several personality characteristics (extraversion) can influence online counseling, help-seeking [43]. Parental behaviors and parental cognitions have also been related to CBT outcomes [44, 45], with CBT treatment for anxious youth successfully changing parent-related variables (parenting, cognitions). Family accommodation for anxiety, defined as the involvement of parents in an anxious youth's efforts to avoid anxiety-provoking activities/situations, is another important factor that should be considered. There is evidence that individual CBT delivered with anxious youths has an effect on parental accommodation [46]; however, much research needs to be conducted regarding this aspect.

5.3. Parent-child relationship

Parent-child relationship is an important variable that has to be considered when designing family interventions, as it is related to treatment outcomes [47]. Relationship quality, family conflicts, and familial dysfunction (for a review of parental involvement in CBT see [48]) should be assessed and considered when designing parental involvement in remote CBT.

5.4. Treatment factors

Several factors related to treatment are: attitudes (expectations, credibility, myths/erroneous information, concerns about remote CBT, treatment preferences), treatment compliance, in- and out-session behaviors [36]. In a study investigating parental attitudes toward cCBT, it becomes apparent that parents are positive about using such interventions to help their anxious child. Except for parental demographic factors, clinical factors, and engagement with technology, knowledge of cCBT treatments is a great predictor of cCBT usage [49]. Indeed, information on e-therapies can influence attitudes toward these treatments [50]. Also, despite the fact that children have high levels of computer usage, they are skeptical when it comes to using cCBT and, generally, their parents exhibit more positive attitudes regarding cCBT treatments [51]. Designing online intervention according to patients' preferences is associated with large reductions in anxiety outcomes, according to a preliminary study in which participants could choose which modules to attend from an iCBT intervention [52].

6. Discussion

The present study aims to present the current evidence on remote CBT interventions delivered for anxious children and adolescents with parental involvement. The systematic review includes eight papers in which remotely delivered interventions have been described. Interventions delivered over the computer (CD-ROM), Internet, telephone, and video were tested against waiting lists, psychoeducation, treatment as usual, and face-to-face CBT. Parental involvement in most of the papers included was directed to teaching parents how to help their children or adolescents deal with situations triggering anxiety and to help their offspring to better acquire the core CBT elements they learn in therapy.

We then proposed a conceptual model in which several factors (parent-, child-, parent-child relationship, treatment-related) were considered in order to design efficient CBT remote interventions for youths with anxiety disorders.

This review should be interpreted considering several limitations. First, we included a small number of studies. Second, the majority of studies included were computerized and Internet CBT interventions. Future studies should test the efficacy of remote CBT interventions in randomized controlled trials. Also, the investigation into mechanisms of change is a major limitation in all these studies, and future research should investigate such factors. We found no study comparing family-based remote CBT with child-based remote CBT, and such studies should be designed in order to have a clear picture of the role of parental involvement in these treatments. What are the implications of involving parents in remote treatments in terms of outcomes, adherence to treatment, and cost-effectiveness? Also, aspects related to when parents should be involved in anxious youths' treatment are unclear, as there is no study investigating the efficacy of only involving parents in exposure sessions, for example, compared with interventions in which parents are involved in all sessions. Parental predictors (demographics, psychopathology, parenting) of successful remote CBT interventions should be investigated in order to guide future developments in this area of research.

http://dx.doi.org/10.5772/65120

In conclusion, our systematic review raises several directions for future research on parental involvement in remote CBT interventions for anxious children and adolescents.

Acknowledgements

Funding for this study was provided by a grant from the Romanian Executive Unit for Financing Education Higher Research, Development and Innovation (the "Effectiveness of an empirically based web platform for anxiety in youths," Grant number PN-II-PT-PCCA-2011-3.1-1500, 81/2012) awarded to dr. Anca Dobrean.

Appendix

Search strategy on PsychInfo

1. "Internet" OR "Internet-based" OR "Internet delivered" OR "web" OR "web based" OR "online" OR "computer-based" OR "computerized" OR "computerized" OR "mobile" OR "eHealth" OR "technology" OR "technology-based" OR "telephone" OR "smart-phone" OR "tablet" OR "app" OR "apps" OR "application" OR "serious game" OR "gamifying" OR "gamified" OR "remote" OR "virtual reality" OR "VR".
2. "anxiety" OR "anxious" OR "social anxiety" OR "social phobia" OR "separation anxiety" OR "phobia" OR "phobic" OR "panic" OR "generalised anxiety" OR "generalized anxiety".
3. "child" OR "children" OR "pediatric" OR "adolescent" OR "teen" OR "youth".
4. "CBT" OR "cognitive behavioral therapy" OR "cognitive behaviour therapy" OR "treatment" OR "intervention" OR "therapy" OR "self-help" OR "program" OR "iCBT" OR "cCBT" OR "eTherapy".
5. "randomised trial" OR "randomized trial" OR "controlled trial" OR "randomization" OR "randomization" OR "random assigned" OR "controlled" OR "trial" OR "RCT".
6. 1 AND 2 AND 3 AND 4 AND 5 AND 6.

Author details

Costina-Ruxandra Păsărelu[1] and Anca Dobrean[2*]

*Address all correspondence to: anca.dobrean@ubbcluj.ro

1 Doctoral School Evidence-Based Assessment and Psychological Interventions, Babeș-Bolyai University, Cluj-Napoca, Romania

2 Department of Clinical Psychology and Psychotherapy, Babeș-Bolyai University, Cluj-Napoca, Romania

References

[1] Polanczyk GV, Salum GA, Sugaya LS, et al. Annual research review: A meta-analysis of the worldwide prevalence of mental disorders in children and adolescents. *J Child Psychol Psychiatry* 2015; 56: 345–365.

[2] Erskine HE, Moffitt TE, Copeland WE, et al. A heavy burden on young minds: The global burden of mental and substance use disorders in children and youth. *Psychol Med* 2015; 45: 1551–1563.

[3] Copeland WE, Shanahan L, Costello E, et al. Childhood and adolescent psychiatric disorders as predictors of young adult disorders. *Arch Gen Psychiatry* 2009; 66: 764–772.

[4] Essau CA, Lewinsohn PM, Olaya B, et al. Anxiety disorders in adolescents and psychosocial outcomes at age 30. *J Affect Disord* 2014; 163: 125–132.

[5] National Institute for Health and Clinical Excellence. Anxiety disorders; 2014. Available from: https://www.nice.org.uk/guidance/qs53/resources/anxiety-disorders-2098725496261 [accessed 29 July 2016].

[6] James AC, James G, Cowdrey FA, et al. Cognitive behavioural therapy for anxiety disorders in children and adolescents. *Cochrane Database Syst Rev* 2015; CD004690.

[7] Mayo-Wilson E, Montgomery P. Media-delivered cognitive behavioural therapy and behavioural therapy (self-help) for anxiety disorders in adults. *Cochrane Database Syst Rev* 2013; CD005330.

[8] Ebert DD, Zarski A-C, Christensen H, et al. Internet and computer-based cognitive behavioral therapy for anxiety and depression in youth: A meta-analysis of randomized controlled outcome trials. *PLoS One* 2015; 10: e0119895.

[9] Kazdin AE, Blase SL. Rebooting psychotherapy research and practice to reduce the burden of mental illness. *Perspect Psychol Sci* 2011; 6: 21–37.

[10] Andersson G. Internet-delivered psychological treatments. *Annu Rev Clin Psychol* 2016; 12: 157–179.

[11] Vigerland S, Thulin U, Ljótsson B, et al. Internet-delivered CBT for children with specific phobia: A pilot study. *Cogn Behav Ther* 2013; 42: 303–314.

[12] Vigerland S, Ljótsson B, Thulin U, et al. Internet-delivered cognitive behavioural therapy for children with anxiety disorders: A randomised controlled trial. *Behav Res Ther* 2016; 76: 47–56.

[13] Tillfors M, Andersson G, Ekselius L, et al. A randomized trial of Internet-delivered treatment for social anxiety disorder in high school students. *Cogn Behav Ther* 2011; 40: 147–157.

http://dx.doi.org/10.5772/65120

[14] Silfvernagel K, Gren-Landell M, Emanuelsson M, et al. Individually tailored internet-based cognitive behavior therapy for adolescents with anxiety disorders: A pilot effectiveness study. *Internet Interventions* 2015; 2: 297–302.

[15] March S, Spence SH, Donovan CL. The efficacy of an internet-based cognitive-behavioral therapy intervention for child anxiety disorders. *J Pediatr Psychol* 2009; 34: 474–487.

[16] Spence SH, Donovan CL, March S, et al. A randomized controlled trial of online versus clinic-based CBT for adolescent anxiety. *J Consult Clin Psychol* 2011; 79: 629–642.

[17] Donovan CL, March S. Online CBT for preschool anxiety disorders: A randomised control trial. *Behav Res Ther* 2014; 58: 24–35.

[18] Morgan AJ, Rapee RM, Tamir E, et al. Preventing anxiety problems in children with Cool Little Kids Online: Study protocol for a randomised controlled trial. *Trials* 2015; 16: 507.

[19] Khanna MS, Kendall PC. Computer-assisted cognitive behavioral therapy for child anxiety: Results of a randomized clinical trial. *J Consult Clin Psychol* 2010; 78: 737–745.

[20] Wuthrich VM, Rapee RM, Cunningham MJ, et al. A randomized controlled trial of the Cool Teens CD-ROM computerized program for adolescent anxiety. *J Am Acad Child Adolesc Psychiatry* 2012; 51: 261–270.

[21] Storch EA, Salloum A, King MA, et al. A randomized controlled trial in community mental health centers of computer-assisted cognitive behavioral therapy versus treatment as usual for children with anxiety. *Depress Anxiety* 2015; 32: 843–852.

[22] Pramana G, Parmanto B, Kendall PC, et al. The SmartCAT: An m-health platform for ecological momentary intervention in child anxiety treatment. *Telemed J E Health* 2014; 20: 419–427.

[23] Whiteside SPH. Mobile device-based applications for childhood anxiety disorders. *J Child Adolesc Psychopharmacol* 2016; 26: 246–251.

[24] Pesky gNATs. Available from: http://peskygnats.com/[accessed 29 July 2016].

[25] Scholten H, Malmberg M, Lobel A, et al. A randomized controlled trial to test the effectiveness of an immersive 3D video game for anxiety prevention among adolescents. *PLoS One* 2016; 11: e0147763.

[26] Brezinka V. Computer games supporting cognitive behaviour therapy in children. *Clin Child Psychol Psychiatry* 2014; 19: 100–110.

[27] Göbl B, Hlavacs H, Hofer J, et al. MindSpace: A Cognitive Behavioral Therapy Game for Treating Anxiety Disorders in Children. In: Chorianopoulos K, Divitini M, Hauge JB, et al. (eds) *Entertainment Computing - ICEC 2015*. Springer International Publishing Switzerland, pp. 463–468.

[28] Pignatiello A, Teshima J, Boydell KM, et al. Child and youth telepsychiatry in rural and remote primary care. *Child Adolesc Psychiatr Clin N Am* 2011; 20: 13–28.

[29] Jones E, Manassis K, Arnold P, et al. Translating cognitive behavioral therapy for anxious youth to rural-community settings via tele-psychiatry. *Community Ment Health J* 2015; 51: 852–856.

[30] Anstiss D, Davies A. 'Reach Out, Rise Up': The efficacy of text messaging in an intervention package for anxiety and depression severity in young people. *Child Youth Serv Rev* 2015; 58: 99–103.

[31] Sarver NW, Beidel DC, Spitalnick JS. The feasibility and acceptability of virtual environments in the treatment of childhood social anxiety disorder. *J Clin Child Adolesc Psychol* 2014; 43: 63–73.

[32] St-Jacques J, Bouchard S, Bélanger C. Is virtual reality effective to motivate and raise interest in phobic children toward therapy? A clinical trial study of *in vivo* with *in virtuo* versus *in vivo* only treatment exposure. *J Clin Psychiatry* 2010; 71: 924–931.

[33] Maldonado JG, Magallón-Neri E, Rus-Calafell M, et al. Virtual reality exposure therapy for school phobia. *Anuario de psicología/The UB Journal of psychology* 2009; 40(2): 223–236.

[34] Manassis K, Lee TC, Bennett K, et al. Types of parental involvement in CBT with anxious youth: A preliminary meta-analysis. *J Consult Clin Psychol* 2014; 82: 1163–1172.

[35] Walczak M, Esbjørn BH, Breinholst S, et al. Parental involvement in cognitive behavior therapy for children with anxiety disorders: 3-year follow-up. *Child Psychiatry Hum Dev*. Epub ahead of print 12 July 2016. doi:10.1007/s10578-016-0671-2.

[36] Pereira AI, Muris P, Mendonça D, et al. Parental involvement in cognitive-behavioral intervention for anxious children: Parents' in-session and out-session activities and their relationship with treatment outcome. *Child Psychiatry Hum Dev* 2015; 47: 113–123.

[37] Moher D, Liberati A, Tetzlaff J, et al. Preferred reporting items for systematic reviews and meta-analyses: The PRISMA statement. *Ann Intern Med* 2009; 151: 264–269, W64.

[38] Higgins JPT, Altman DG, Gøtzsche PC, et al. The Cochrane Collaboration's tool for assessing risk of bias in randomised trials. *BMJ* 2011; 343: d5928.

[39] McGrath PJ, Lingley-Pottie P, Thurston C, et al. Telephone-based mental health interventions for child disruptive behavior or anxiety disorders: Randomized trials and overall analysis. *J Am Acad Child Adolesc Psychiatry* 2011; 50: 1162–1172.

[40] Infantino A, Donovan CL, March S. A randomized controlled trial of an audio-based treatment program for child anxiety disorders. *Behav Res Ther* 2016; 79: 35–45.

[41] Liber JM, van Widenfelt BM, Goedhart AW, et al. Parenting and parental anxiety and depression as predictors of treatment outcome for childhood anxiety disorders: Has the role of fathers been underestimated? *J Clin Child Adolesc Psychol* 2008; 37: 747–758.

[42] Schleider JL, Ginsburg GS, Keeton CP, et al. Parental psychopathology and treatment outcome for anxious youth: Roles of family functioning and caregiver strain. *J Consult Clin Psychol* 2015; 83: 213–224.

[43] Tsan JY, Day SX. Personality and gender as predictors of online counseling use. *J Technol Human Serv* 2007; 25: 39–55.

[44] Esbjørn BH, Sømhovd MJ, Nielsen SK, et al. Parental changes after involvement in their anxious child's cognitive behavior therapy. *J Anxiety Disord* 2014; 28: 664–670.

[45] Festen H, Hartman CA, Hogendoorn S, et al. Temperament and parenting predicting anxiety change in cognitive behavioral therapy: The role of mothers, fathers, and children. *J Anxiety Disord* 2013; 27: 289–297.

[46] Kagan ER, Peterman JS, Carper MM, et al. Accommodation and treatment of anxious youth. *Depress Anxiety*. Epub ahead of print 23 May 2016. doi:10.1002/da.22520.

[47] Crawford AM, Manassis K. Familial predictors of treatment outcome in childhood anxiety disorders. *J Am Acad Child Adolesc Psychiatry* 2001; 40: 1182–1189.

[48] Breinholst S, Esbjørn BH, Reinholdt-Dunne ML, et al. CBT for the treatment of child anxiety disorders: A review of why parental involvement has not enhanced outcomes. *J Anxiety Disord* 2012; 26: 416–424.

[49] Sweeney GM, Donovan CL, March S, et al. Logging into therapy: Parent attitudes and intentions to use computer-based therapies for youth mental health. *Internet Interventions* 2015; 2: 437–445.

[50] Casey LM, Joy A, Clough BA. The impact of information on attitudes toward e-mental health services. *Cyberpsychol Behav Soc Netw* 2013; 16: 593–598.

[51] Stallard P, Velleman S, Richardson T. Computer use and attitudes towards computerised therapy amongst young people and parents attending child and adolescent mental health services. *Child Adolesc Ment Health* 2010; 15: 80–84.

[52] Andersson G, Estling F, Jakobsson E, et al. Can the patient decide which modules to endorse? An open trial of tailored internet treatment of anxiety disorders. *Cogn Behav Ther* 2011; 40: 57–64.

Chapter 7

Impact of Social Media on Social Anxiety: A Systematic Review

Anca Dobrean and Costina-Ruxandra Păsărelu

Additional information is available at the end of the chapter

http://dx.doi.org/10.5772/65188

Abstract

Introduction: Online social networking sites are being used all around the world. However, only recently researchers have started to investigate their relationship with mental health. Evidence coming from literature suggests that they have both advantages and disadvantages for individuals. The aim of this study is to critically review the existent research conducted on the relationship between online social networking and social anxiety.

Method: According to PRISMA guidelines, comprehensive systematic searches of electronic databases were conducted (PsychInfo, Cochrane, PubMed, Scopus, Web of Science). Terms related to online social networking were combined with social anxiety terms. After identifying relevant papers, the relationship between social media networking and social anxiety as well as limitations of each study were presented. All of the papers included were cross-sectional, relying on self-report assessments.

Conclusion: There are several papers published on the relationship between social media and social anxiety; however, more research needs to be done in order to clarify the influence of one variable over the other. Rigorous experimental studies need to be conducted with different time-point assessments, and bidirectionality should be investigated. Important mediators and moderators have to be considered in order to explain the relationship between social media networking and social anxiety.

Keywords: social media, social anxiety, online social networking, Facebook

1. Introduction

Social networking sites have become part of twenty-first century people's lives. From all online social networking sites, Facebook is the most widely used. According to a recent report published by Facebook regarding data for the second quarter of 2016, there is a considerable increase (17% in comparison with last year) in the number of both daily and monthly active users, with 1.13 billion daily active Facebook users for June 2016 and approximately 1.71 billion monthly active Facebook users for 30 June 2016 [1].

2. Online social networking

2.1. Benefits of online social networking

Online social networking sites may have a useful role in mental health research (for a comprehensive review of the use of Facebook in social science see Ref. [2]). For instance, Facebook, the most widespread form of online social networking site, can easily be used as a great clinical research tool, as it can provide the recruitment of patients and involve them in programmes [3]. Also, it can be used to guide interventions and in treatment monitoring (e.g. in physical activity or overweight programs [4, 5]). So far, several studies have investigated the role of social media in people with serious illness, and research shows that it is a promising approach in patients with schizophrenia [6].

Social media is a valuable resource for receiving peer-to-peer support. Facebook, Twitter or YouTube can be used by people with several conditions in order to find support or advice from others and to share personal experiences. Naslund et al. [7] proposed a model that illustrated potential benefits for people with serious mental illness, which they can have in an online community on online social networking sites (overcome stigma, seek professional help, receive adequate treatment).

Regarding mental health interventions delivered via Facebook, research is only starting to emerge. In an online randomized controlled experiment, a depression awareness campaign delivered via Facebook for adults was associated with enhanced mental health literacy [8]. Facebook allows for establishing and maintaining connections with others, and studies show that there is a positive relation between Facebook use and social capital, with greater advantages for people with low self-esteem and life satisfaction [9]. According to a review on the applications of social media in medicine and healthcare services, it seems that social networking is a promising approach. However, much uncertainty exists in terms of ethics and safety [10]. Furthermore, social media seemed to be an effective method to promote health-related behaviours [11].

There are studies showing that online social networking sites are important factors in youths' social lives, as in a large sample ($N = 3.068$) of adolescents (aged 11–14), there were significant positive associations between the use of online social networking sites and several friendship-related variables (friendship quality, face-to-face interaction, bridging/bonding social capital) [12].

Online social networking sites can have benefits in terms of cognitive abilities for older healthy adults [13], as it was shown in a study involving older adults (mean age comprised between 78 and 80), in which participants were randomly assigned into three groups: Facebook training, online diary website and waiting list. The Facebook intervention was delivered over 8 weeks and older adults received weekly training in how to use Facebook.

2.2. Disadvantages of online social networking

Recent research has associated social media networking with several negative outcomes, both in adolescent populations and in adults.

2.2.1. Adolescent literature

According to a report published by the Pew Research Center, adolescents are avid users of online social networking sites, with approximately 71% of them using more than one online social networking site [14] in 2015, and the most widely used online social networking site was Facebook (41%). Frequency of using social networking sites expressed as the amount of time spent on online social networking sites was associated with mental health problems in children and adolescents [15]. Namely, the same study shows that spending more than 2 h per day on online social networking sites is associated with higher psychological distress, poor self-rated mental health, suicidal ideation and an unmet need for mental health support. In another study conducted on a large sample of adolescents (N = 5.126), there were significant positive associations between the use of online social networking sites, psychological distress, suicide ideation and suicide attempts [16]. In this study, cyberbullying victimization served as a full mediator in the relationship between the use of online social networking sites and psychological distress/suicide attempts and acted as a partial mediator in the relationship between the use of online social networking sites and suicidal ideation.

"Facebook depression" [17] is a term introduced in a report of the American Academy of Pediatrics to describe the impact of social media on youths' mental health, according to which depression arises as a consequence of youths spending a large amount of time on social media [18]. When it comes to mechanisms explaining why the use of online social networking sites is associated with negative mental health outcomes, one study shows that negative comparison on Facebook is related with adolescents' life satisfaction [19]; negative comparison on Facebook predicts life satisfaction, but the opposite relationship is also significant.

Research regarding problematic online social networking sites in adolescents is scarce, and very few studies have investigated the relationship between Facebook addiction and youths' mental health problems. In a study conducted with adolescents, both personality traits and social influence processes emerged as significant predictors of problematic Facebook use and Facebook use frequency. Problematic Facebook use was predicted by emotional stability, extraversion, conscientiousness and norms, while the frequency of Facebook use was predicted by gender, group norms and social identity [20].

2.2.2. Adult studies

With the rapid emergence of technology, the term "iDisorders" was introduced to define mental health problems related to technology usage. In a meta-analysis including eighteen papers on the relationship between Facebook use and loneliness, it resulted that the two variables were significantly associated and that loneliness predicted Facebook use and not the other way around [21]. Facebook use had a significant contribution to mental health, with Facebook use, impression management and friends predicting mood disorders, with different contributions across disorders (having more friends is negatively associated with major depression and dysthymia, while positively predicting mania) [22]. Social media, operationalized as the total time spent per day, visits per day and global frequency of visits on social media, was associated with depression in a large sample of U.S. young adults [23]. Facebook intrusiveness was correlated with obsessive-compulsive disorder severity and with obsessive beliefs in a sample of young adults (aged 18–24) [24].

There is ample literature on the role of online social networking sites, and Facebook in particular, in mental health outcomes. However, there is high heterogeneity in what online social networking sites and Facebook mean. Facebook is conceptualized and measured differently across studies, given that while several studies investigate Facebook use, others investigate either the number of friends, Facebook disclosure, Facebook activities, Facebook motives or Facebook addiction/intrusiveness/problematic use.

Despite the fact that no diagnostic manual recognizes Facebook addiction as a diagnostic category, more and more research focuses on pathological uses of online social networking sites, and Facebook in particular (see Ref. [25] for a complex review of online social networking sites addiction). Online social networking sites addiction is related to deficits in emotion regulation and susceptibility to drug and alcohol addictions [26], somatic symptoms, anxiety, insomnia, social dysfunction and depression [27].

3. Social media and social anxiety

Online social networking sites, and Facebook in particular, can offer great opportunities, in terms of interaction with others, for individuals with social anxiety. Two different hypotheses have been proposed in order to explain why socially anxious individuals use computers [28], which can easily be applicable to online social networking sites. The first hypothesis proposed is the social *compensation hypothesis* [29], according to which individuals use online social networking sites in order to compensate for deficits in social skills or discomfort in face-to-face situations. The second theory, opposed to the first, is called the social enhancement hypothesis. According to this hypothesis, socially skilled individuals use online social networking sites to find additional opportunities to interact with others. In the existent literature on the relationship between online social networking sites and social anxiety, there is mixed evidence regarding the two hypotheses, as both have received support.

In an experimental study, Rauch et al. [30] investigated whether previous exposure to Facebook attenuates or augments physiological arousal during face-to-face interaction in a sample

http://dx.doi.org/10.5772/65188

composed exclusively of adult female students. Participants were randomly assigned to four groups: Facebook exposure only, face-to-face exposure only, Facebook exposure first plus face-to-face exposure second and face-to-face exposure first plus Facebook exposure second. The results indicated that participants who only had seen the stimulus on Facebook had less physiological arousal than participants who had seen the stimulus face-to-face. However, regarding the combined conditions, physiological arousal was higher when seeing the person face-to-face when this preceded seeing the person on Facebook. Social anxiety moderated the relationship between physiological arousal and Facebook exposure so that exposing an individual to a stimulus person on Facebook is related with increased arousal in a face-to-face meeting with that person, and this, moreover, appears in highly social anxious persons.

No systematic approach has been conducted so far investigating the relationship between social anxiety and online social networking sites, or Facebook in particular, despite growing evidence on how socially individuals use online social networking sites. Therefore, the main aim of the current paper is to offer a comprehensive review of the state of the art of existent studies on the relationship between online social networking sites and social anxiety.

4. Methods

The current systematic review has been conducted in accordance with the Preferred Reporting Items for Systematic Reviews and Meta-Analyses (PRISMA) statement [31].

4.1. Inclusion and exclusion criteria

In order to be included, studies had to:

(a) involve participants with social anxiety symptoms/disorder;

(b) include online social networking assessments;

(c) include social anxiety assessments;

(d) be published in peer-reviewed journals;

(e) be published in English.

There were no age constraint criteria applied and no year of publication limitation.

Studies were excluded if there was no assessment of social networking and social anxiety, and they were not published in English.

4.2. Search strategy

Electronic databases (PsychInfo, PubMed, Scopus, Web of Science) were consulted by two independent assessors up to 20th June 2016. Several terms related to social networking were combined with terms related to social anxiety. No additional records were identified through the manual search in the references of the articles published and meta-analyses related to the subject. The reference lists of relevant papers and meta-analyses related to the subject were screened in order to identify potentially relevant articles. We repeated the search on 25th July 2016 to double-check and identify whether new articles had been published since the first search. The search terms were as follows:

1. Social network* OR Facebook OR LinkedIn OR Tweeter OR Instagram OR Youtube* OR Myspace OR Social Network Sites OR social media*

2. Social phobia OR social anx* OR social phobi* OR social* anxious* OR generalised social anx* OR generalized social anx*

3. 1 AND 2

5. Results

5.1. Literature search

The flowchart describing the inclusion of studies is described in **Figure 1**. The literature search yielded 673 articles, of which 98 were duplicates. After removing duplicates, the titles and abstracts of 575 articles were screened. Thirty-eight relevant papers were selected for full-text reading. Of these papers, 18 were excluded based on the following reasons: no social anxiety assessments (n = 20), not peer-reviewed articles (n = 2), not published in English (n = 1), no measure on social networking (n = 1), and theoretical reviews or meta-analyses (n = 2). This left a total number of 20 articles to be included in the systematic review.

5.2. Information extracted

The following information was extracted from each of the studies included:

(a) authors and year of publication;

(b) sample characteristics (number of participants, mean age, age range, proportion of female participants, type of sample);

(c) amount of time spent using social networking sites per day;

(d) online social networking conceptualization;

(e) concepts assessed in the study;

(f) information regarding the association between online social networking and social anxiety.

http://dx.doi.org/10.5772/65188

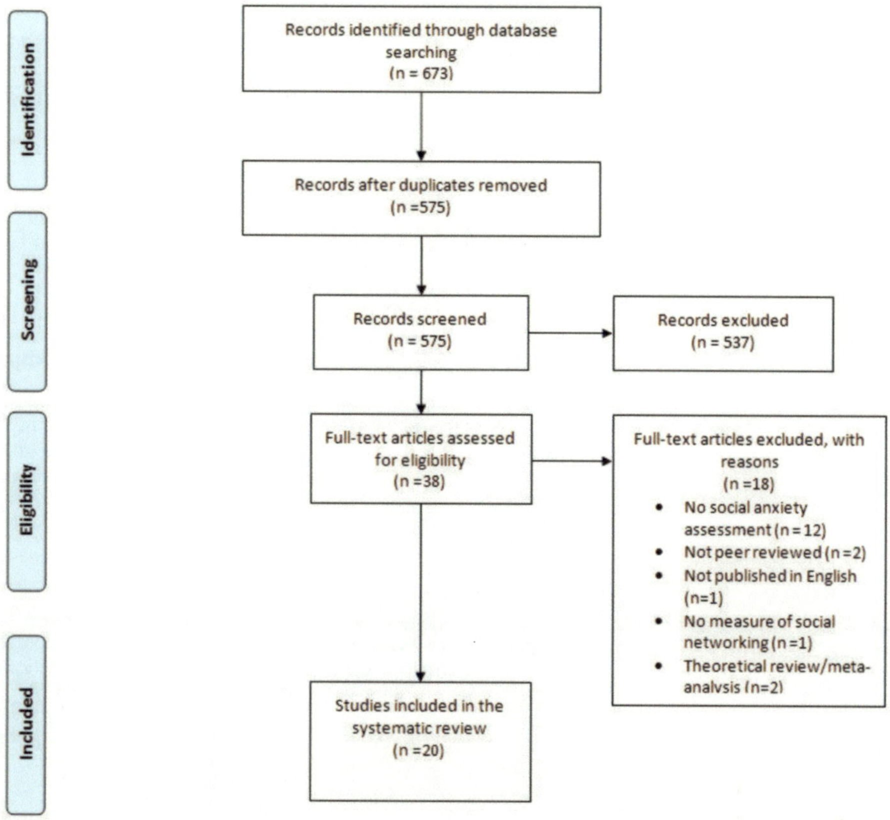

Figure 1. Flow diagram of study selection process.

5.3. Characteristics of the studies included

A summary of the characteristics of the studies included (number of participants, age, online social networking sites conceptualization, primary results) is described in **Table 1**.

Study name	*N* (% F)	M_{age} (SD) Age range	Sample	Time spent in social media per day (SD)	SNSs construct	Measures	Primary results
Aladwani and Almarzouq [45]	407 (54%)	20.04 (1.16)	Students	1.86 h (0.99)	Compulsive SM use	Interaction anxiousness Self-esteem Problematic learning Media complementarity	Significant positive associations between compulsive SM use and SA $r = 0.26, p < 0.05$
Antheunis et al. [12]	3.068 (53.7%)	13.46 (0.56) 11–14	Adolescents	7.27 h (8.54)	SNSs use intensity	Social anxiety Quality of friendships	Significant negative associations between SNS and SA

Study name	N (% F)	M_{age} (SD) Age range	Sample	Time spent in social media per day (SD)	SNSs construct	Measures	Primary results
						Bridging social capital Bonding social capital Offline interaction with friends	$r = -0.08$, $p < 0.001$
Bodroža and Jovanović [32]	804 1st phase 445 (79.1%) 2nd phase 359 (79.4%)	26.95 (6.35) 15–62 21.29 (2.96) 18–44	Students and general population	Combined sample: ½ of the sample uses FB 1–3 h/day, 27% <1 h, 15% 3–5 h and 8% >5 h Student sample: FB 1–3 h/day, 35% <1 h, 11% 3–5 h, 3% >5 h a day.	Psycho-social aspects of Facebook use (PSAFU): Compensation Self-presentation Socialization FB addiction Virtual self	Social anxiety personality Sensation seeking	Significant positive associations between FB and SA r between 0.11 and 0.58, $p < 0.001$ and 0.05 in both samples Significant positive associations between FBA and SA $r = 0.19$, $p < 0.001$ (FB users sample) $r = 0.17$, $p < 0.001$ (Student sample)
Casale and Fioravanti [46]	400 (51.8%)	22.45 (2.09)	Students	Not reported	Generalized problematic Internet use	Social anxiety Satisfaction with the needs through SNS: -The need to belong-The need for self-presentation -The need for assertiveness	Significant positive associations between problematic Internet use and SA $r = 0.22$, $p < 0.001$
Clayton et al. [39]	229 (74.3%)	18.19 (0.43) 18–21	Students	30 min to 1 h	Emotional connectedness to FB Facebook connection strategies	Anxiousness-social anxiety subscale loneliness Alcohol use Marijuana use	Significant positive associations between emotional connectedness to Facebook and anxiousness $r = 0.18$, $p < 0.001$ Significant positive associations between Facebook connection strategies and anxiousness $r = 0.23$, $p < 0.001$
Davidson and Farquhar [49]	336 (70%)	Not reported	Students	Not reported	Facebook intensity Facebook anxiety	Social anxiety number of unique groups Role conflict Self-monitoring Religious activities	Non-significant positive associations between FB Intensity and SA $r = 0.05$, $p = ns$ Significant positive associations between FB anxiety and SA $r = 0.66$, $p < 0.001$
Fernandez et al. [29]	62 (63%)	19 (1.05)	Students	Not reported	Facebook usage	Social anxiety Depression Personality	Significant negative associations between FB friends and SA $r = -0.45$, $p < 0.001$

http://dx.doi.org/10.5772/65188

Study name	N (% F)	M_{age} (SD) Age range	Sample	Time spent in social media per day (SD)	SNSs construct	Measures	Primary results
							Significant positive associations between FB usage and SA *r* between 0.27 and 0.50, $p < 0.001$ (i.e., number of lines in about me, number of TV shows, number of music interests) Non-significant associations between SA and number of activities, self-reported time spent on FB, No. of status updates, No. of posts by friends, and no of posts by self.
Green et al. [43]	306 (65.69%)	20.52 (1.45)	Online participants from all around the world	2.44 h (2.82)/day	FB self-disclosure (public and private)	Social anxiety Controllability Reduced cues Disinhibition Offline self-disclosure	Non-significant associations between FB self-disclosure public and SA $r = 0.05$, p = ns Non-significant associations between FB self-disclosure private and SA $r = 0.00$, p = ns Non-significant associations between time spent on FB and SA $b = 0.11$, p = ns
große Deters et al. [33]	1st study 153 (60.78%) 2nd study 209 (88.52%)	1st study 20.18 (3.24) 2nd study 23.50 (3.24)	1st study Students 2nd study General population	Not reported	Status updates, No. of likes received per status update No. of individual commenters per status update	Social anxiety personality (extraversion)	SA did not predict no. status updates, No. of likes received per status update, No. of individual commenters per status update p = ns
Hong et al. [40]	230 (31.7%)	Not reported	Students	34.3% spent < 1 h; 30.9% spent 1–2 h; 17.4% 2–4 h; 17.4% >4 h	Continuance intention to interact with others on FB Online social anxiety	General social anxiety Belief in dangerous virtual communities	Significant negative associations between FB continuance intention and SA $r = -0.48$, $p < 0.001$ Significant positive associations between OSA and SA $r = 0.38$, $p < 0.001$
Indian and Grieve [41]	299 (85.95%)	28.35 (10.88)	Recruited from Facebook	30–60 min, 14% with daily usage >3 h	Facebook social support items	Social anxiety Offline social support Subjective well-being	No difference in perceptions of Facebook social support in high and low socially anxious, $t(297) = 0.14$, $p = 0.89$ FB social support explained a significant amount of additional

Study name	N (% F)	M_{age} (SD) Age range	Sample	Time spent in social media per day (SD)	SNSs construct	Measures	Primary results
							variance in subjective well-being over and above offline social support for the high social anxiety group
Landoll et al. [34]	1st study: 216 (63%) 2nd study: 214 (54%)	19.06 (1.28) 15.72 (1.22)	Adolescents and young adults	Not reported	Aversive social networking peer experiences	Social anxiety Internet use Social Peer victimization Depression	Significant positive associations between negative SNSs experiences and SA $r = 0.37, p < 0.001$
Lee [47]	304 (56%)	22.45 (6.10) 17–55	College students	Not reported	Facebook addiction	Social anxiety Smartphone addiction Personality traits Multitasking	Significant positive associations between FBA and SA $r = 0.25, p < 0.01$
Lee-Won [35]	243 (71.6%)	19.69 (1.12)	College students	Assessed but not reported	Amount of FB use Problematic Facebook use	Social anxiety Personality Need for social assurance	Non-significant negative associations between FB use and SA $r = -0.03, p = \text{ns}$ Significant positive associations between FBA and SA $r = 0.45, p < 0.001$
Liu et al. [44]	780 (50.9%)	13.94 (0.90) 13–18	Adolescents	Not reported	Personally identifiable information (PII) disclosure	Social anxiety Privacy concern Parental mediation Narcissism	Non-significant negative associations between PII and SA $r = -0.07, p = \text{ns}$ SA indirectly decreases PII disclosure by increasing privacy concern.
McCord et al. [38]	216 (85.64 %)	32.2 (12.43) 18–69	31.5% undergraduate students, 12.0% graduate students, and 55.6% were not students	Not reported	FB use FB social interaction anxiety	Social anxiety	Non-significant associations between FB use and SA $r = 0.01, p = \text{ns}$ Significant positive associations between FB SA and SA $r = 0.64, p < 0.001$
Ramirez et al. [42]	244 (51%)	34.59 (11.19)	Facebook users	Not reported	FB reconnection	Social anxiety Information seeking Uncertainty predicted outcome value Sociability	Marginal significant negative associations between SA and FB reconnect $b = 0.14, p = 0.05$
Rizvi [48]	150 (75%)	18–27	Different educational institution	46% <30 min 32.7% 30 min to 1 h 11.3% 1–2 h 10% > 2 h	Excessive FB use (log in)	Social anxiety Personality	Non-significant negative associations between excessive FB use and SA $r = -0.02, p = \text{ns}$
Shaw et al. [36]	75 (55.2%)	19.2 (1.27)	Students	2.04 (1.13) Range 1–5	FB time passive FB use	Rumination	Significant positive associations between

Study name	N (% F)	M_{age} (SD) Age range	Sample	Time spent in social media per day (SD)	SNSs construct	Measures	Primary results
					FB content production Interactive FB communication	Depressive symptoms Mood and anxiety symptoms Content production Interactive communication Social anxiety	time spent on FB and SA $r = 0.33$, $p < 0.01$ Significant positive associations between passive FB use and SA $r = 0.32$, $p < 0.01$ Non-significant associations between content production on FB $r = 0.23$, $p =$ ns Non-significant associations between interactive FB communication and SA $r = 0.21$, $p =$ ns
Weidman and Levinson [37]	77 (77%)	18.91 (1.05)	Students	Not reported	Facebook profile (no. friends, no. photographs, no. videos, photo albums, relationship status, people in profile picture, status update, quote) Social activity Social inactivity Close relationship quality Self-disclosure	Social interaction anxiety	Significant negative associations between no. friends and SA $r = -0.21$, $p < 0.05$ Significant negative associations between FB quote length and SA $r = -0.38$, $p < 0.05$

Note: N = number of participants; $\%F$ = percentage of female participants; M_{age} = mean age; SD = standard deviation; SNSs = social networking site; SM = social media; SA= Social anxiety; FB= Facebook; FBA= Facebook addiction; OSA = online social anxiety; ns= non-significant; no. = number; h = hours; min = minutes.

Table 1. Characteristics of the included studies.

We included 20 papers in the systematic review, of which three papers [32–34] consisted of two studies/phases each. Most of the studies included were conducted with adults (n = 14), recruited either from university (students), or from the general populations, while several papers were conducted with adolescents (n = 3), and some studies did not report the age of the participants (n = 2). Nine of the papers included reported the time spent on Facebook; one study also assessed this indicator but did not report it, and the rest of the studies did not include reports on the time spent on Facebook. Social media was operationalized differently in the papers included and while few studies reported social media variables, most of the papers included referred specifically to Facebook.

With respect to the manner in which online social networking sites or Facebook were conceptualized in these studies, we have encountered several approaches:

- *Facebook use* (n = 7), which in turn can be conceptualized as the amount of time spent on online social networking sites, passive use of Facebook or using interactive Facebook features (updates, comments, likes);
- *Psychosocial aspects of Facebook use* (n = 6), which can also mean Facebook connection strategies, Facebook continuance intention, Facebook reconnection decision, Facebook social support, aversive Facebook experiences or cybervictimization;
- *Personal disclosure on Facebook* (n = 3), which can also mean private or public disclosure, personal identifiable information (profile information and privacy settings) and status/quote update/length;
- *Problematic use of online social networking sites* (n = 6), which can be found as Internet addiction, compulsive social media use, Facebook addiction or Facebook intrusiveness;
- *Facebook anxiety* (n = 3), with different terms such as Facebook-specific anxiety, Facebook interaction anxiety or online anxiety.

Most of the papers included found a significant association between social networking (Facebook) and social anxiety (n = 16), while the rest of the studies included found no significant relationship between social media and social anxiety.

5.4. Online social networking and social anxiety

5.4.1. Usage of online social networking sites and social anxiety

There are mixed findings regarding the relationship between online social networking and social anxiety. On the one hand, there are studies reporting significant associations between the two variables. For example, Antheunis et al. [12] showed that Facebook use was negatively associated with social anxiety in a large sample of adolescents. Lee-Won [35] found evidence that Facebook use, defined as the amount of time participants spend on Facebook on an average day, was associated with social anxiety. Shaw et al. [36] found a significant relationship between the time spent on Facebook and social anxiety in a sample of undergraduate students. The same study found a significant association between passive Facebook use, defined as passive content consumption (e.g. passively viewing one's Facebook profile) and social anxiety. The relationship between passive Facebook use and social anxiety was significant even after checking for depressive and anxiety symptoms. Weidman and Levinson [37] considered both offline (self-reported indicators) and online indicators (using the profiles of coders for Facebook) of social anxiety. Results on self-reported social anxiety and objective Facebook use indicated significant negative relationships between the two variables, with number of friends, number of photographs and quotes length being negatively related to social anxiety. Regarding the rated social anxiety symptoms observed and objective Facebook signs, there were significant negative associations between social activity composite (number of friends, photographs, videos, photo albums), the number of people in profile pictures and social anxiety.

http://dx.doi.org/10.5772/65188

On the other hand, there are several research papers that show no association between Facebook use and social anxiety. große Deters et al. [33] found no significant relationship between social anxiety and different parameters of Facebook use (status updates, number of likes or individual commenters per like), and none of the prediction models investigated, in which social anxiety was used as a predictor and Facebook as a criterion, were significant. McCord et al. [38] found no significant relationship between general Facebook use, defined as the frequency of using the socially interactive features of Facebook (sending messages, using chat, writing on events/friends walls, sending friend requests, posting comments, updating status) and social anxiety. Fernandez et al. [29] did not find a positive association between the frequency of Facebook usage, operationalized as time spent on Facebook, status updates, number of posts by friends/self and social anxiety, in a sample of students. In this study, several independent coders rated each participant's level of social anxiety based on viewing their Facebook profiles and associated it with participants' self-rated social anxiety symptoms. Social anxiety was not associated with an increased frequency of Facebook use, either based on self-report data or based on coders' reports. However, there were positive associations between profile information sections and social anxiety. The number of activities was not associated with social anxiety. Using multiple regression, the results indicated that social anxiety had a unique contribution, apart from depression and neuroticism, to profile information (excepting the number of movies and activities).

5.4.2. Psychosocial aspects of Facebook use and social anxiety

Bodroža and Jovanović [32] constructed a scale in order to measure several psychosocial aspects of Facebook use. Five factors emerged for the structure of the scale, namely: compensation (use Facebook in order to compensate for personal insecurity/inadequacy), self-presentation (concern related to impression on others), socialization (striving to acquaint new friends/intimate partners), Facebook addiction and virtual self (Facebook adequately represents one's personality). Social anxiety had positive associations with each of these factors. Clayton et al. [39] found a positive relationship between anxiousness (public and private self-consciousness, social anxiety), emotional connectedness to Facebook (integration of Facebook in one's daily life) and Facebook connection strategies (reasons to use Facebook, use Facebook to learn more information about people, to maintain current relationships). Anxiousness was a significant predictor of both emotional connectedness to Facebook and Facebook connection strategies. Hong et al. [40] investigated the relationship between the continuance intention to interact with other people on Facebook and social anxiety and found a significant negative relationship between them. Social anxiety predicted the continuance intention to interact with other people on Facebook and, together with online social anxiety, accounted for 44% of the continuance intention. Indian and Grieve [41] found no differences between low and high socially anxious individuals in Facebook social support. There were significant association in both groups between Facebook social support and subjective well-being, as well as between Facebook social support and offline social support. Finally, Facebook social support was a significant individual predictor of subjective well-being for highly social anxious individuals. Landoll et al. [34] developed an instrument that assesses negative peer experiences on social networking sites. There were significant associations between negative online peer experiences and social anxiety and depression. Ramirez et al. [42] investigated factors related to the decision to reconnect on Facebook with a past contact in a sample group of adult participants. Using hierarchical logistic regression, they found several predictors that accounted for almost 40% of the variance of the decision to reconnect on Facebook, in which social anxiety was also included and has a marginally negative contribution to it. There were other factors (e.g. relational factors and information seeking factors) that have a greater contribution to the decision to reconnect than social anxiety.

5.4.3. Personal disclosure on Facebook and social anxiety

There was no significant evidence stemming from the papers included of an association between Facebook disclosure and social anxiety. Green et al. [43] found no relationship between Facebook self-disclosure (public and private) and social anxiety. Two hypothesized pathways from social anxiety to Facebook self-disclosure considered online attributes (reduced cues, controllability) and feelings of disinhibition. Results showed that the model accounted for 23% of the variance in Facebook private self-disclosure and 7% of the variance in Facebook public self-disclosure. Liu et al. [44] did not find any relationship between Facebook disclosure and social anxiety in a sample group of adolescents aged between 13 and 18. Disclosure in this study referred to personally identifiable information, a composite score computed by adding a measure for attitudinal information disclosure (profile information on Facebook page and personal photographs) and a scale assessing behavioural information disclosure (items actually posted on Facebook and privacy regarding photographs). There was only an indirect relationship between social anxiety and Facebook disclosure through the role of privacy concern, defined as concern related to the security of the personal information presented online. Namely, by increasing privacy concern, social anxiety indirectly decreases disclosure. Using a path analysis approach, parental mediation had only an indirect effect on disclosure through privacy concern, and no direct effect. Self-disclosure, defined as the presence or absence of status update/quote and length of status update/quote, was also considered in Weidman and Levinson [37] as a measure of social inactivity. Status update was unrelated to social anxiety in both self-reported and observer-rated social anxiety and objective Facebook signs, while the length of the quote was related to social anxiety only in self-reported and not in observer-rated social anxiety.

5.4.4. Problematic social media use/Facebook addiction and social anxiety

Most of the papers included sustained the relationship between problematic use of online social networking sites or online social networking sites addiction and social anxiety. For instance, Aladwani and Almarzouq [45] found a significant positive association between interaction anxiousness and compulsive social media used, which finally has a significant effect on learning outcomes. Bodroža and Jovanović [32] found a significant relationship between social addiction and social anxiety both in a sample of Facebook users recruited online and in a sample of students. Casale and Fioravanti [46] considered the Self-Determination Theory in the context of Facebook use and proposed that the satisfaction for unmet needs through social networking sites should be a viable candidate in the development of problematic use of Internet communicative services. Results showed that social anxiety had a significant effect on generally problematic Internet use for both women and men, in a sample group of undergraduate students. They also considered several needs for using social networking sites, such as the need to belong, self-presentation and assertiveness. Problematic use of Internet communicative services was significantly associated with all the three needs considered. However, only the need for self-presentation was a significant mediator in the relationship between social anxiety and problematic Internet use only in the case of males. Lee [47] investigated the relationship between Facebook addiction and social anxiety in a sample group of African American students and found a significant positive relationship between the two variables. Using hierarchical multiple regression, the results showed that approximately 19% of the variance of Facebook addiction was explained by eight predictors, of which age, social interaction anxiety and multitasking were the most related to Facebook addiction. Finally, Facebook addiction was significantly associated with smartphone addiction. Lee-Won [35] found a significant association between problematic Facebook use and social anxiety, with the latter being a noteworthy predictor for problematic Facebook use. Furthermore, the results indicated that a significant moderator in the relationship between problematic Facebook use and social anxiety was the need for social assurance. Rizi [48] found no relationship between social anxiety and excessive Facebook use in a sample of 150 young adults; there were no differences in Facebook use according to high vs. low social anxiety levels. However, the majority of the participants from this study logged on to their Facebook accounts few times a day (47%) and spent less than 30 min on Facebook (46%). The only parameter regarding Facebook usage that was considered in this study for the association with social anxiety was the frequency with which participants log onto Facebook, with no other variables considered.

5.4.5. Facebook anxiety and social anxiety

Three of the studies included that assessed online social networking sites anxiety indicated significant positive relationships with social anxiety. Davidson and Farquhar [49] investigated the relationship between social anxiety, religion and Facebook. They developed a Facebook-specific anxiety scale, an adaptation of a social anxiety scale related to Facebook and found a strong association with social anxiety. Online social anxiety and social anxiety were positively associated in Hong et al. [40], with general social anxiety being a significant predictor of online social anxiety. McCord et al. [38] used a Facebook social interaction scale and found a significant association with social anxiety in an adult sample group. Using multiple regression results indicated that social anxiety and anxiety on Facebook predicted Facebook social use. There was a significant interaction between social anxiety and Facebook social anxiety, which is in line with the social compensation theory. According to this theory, highly socially anxious individuals tend to use Facebook in order to compensate for the discomfort associated with face-to-face communication. A second regression model was also significant, with anxiety on Facebook and social Facebook use predicting social anxiety; again, there was a significant interaction between the two predictors.

6. Discussion

Online social networking sites are widespread means through which people can interact with others. There are many advantages of online social networking sites; however, recent research focuses mainly on the negative impact they have on mental health. Terms like "iDisorders", "Facebook depression" and "Facebook addiction" abound in the literature on online social networking sites and mental health, proposing new disorders that are determined by the use of online social networking sites/Facebook. Moreover, their particularities in terms of manifestations make some advocate their inclusion in current diagnostic nosologies.

The present review aims to synthetically present the existing literature on the relationship between social anxiety and social networking. Twenty studies met our inclusion criteria. Most of these studies referred to a particular online social networking site, Facebook, and assessed different related parameters. Overall, most of the papers reported significant associations between online social networking and social anxiety, with four papers reporting no relationship between the two variables. There were mixed results on the relationship between Facebook use and social anxiety, as there were both studies sustaining a significant positive association between the two variables and those that found no association. Several psychosocial aspects of Facebook use were significantly related, both positively and negatively, to social anxiety. There was no relationship between Facebook disclosure and social anxiety in any of the papers included that assessed this construct. With one exception, there was a significant positive relationship between online social networking sites addiction/ problematic use and social anxiety. There was evidence of a significant association between Facebook anxiety and social anxiety.

This results should be interpreted in the light of the fact that there was a high heterogeneity of Facebook conceptualizations, as in several studies, the amount of time spent on Facebook was assessed, while other studies assessed Facebook/ Internet/ social networking problematic use/intrusiveness/addiction, or psychosocial aspects related to Facebook use, or Facebook disclosure, and in fewer studies, Facebook anxiety was considered. Therefore, one cannot conclude that online social networking sites have a negative or a positive effect on social anxiety without carefully taking into consideration what they mean. An important limitation of the current review was the fact that although we aimed to investigate the relationship between social media and social anxiety, most of the papers included referred only to social networking sites such as Facebook and we cannot generate findings pertaining to other social networking sites, such as Instagram, Tweeter, Youtube, LinkedIn or Myspace. Future studies should also investigate the role of these other social networking sites in mental health problems and in social anxiety per se. Another limitation of this paper refers to the sample groups. As most of the participants included were students, this does not enable us to extend our to other populations. Involving more heterogeneous samples of participants, adolescents, young and older adults can help us to find significant moderators in the relationship between social networking and social anxiety. As all the studies included were

http://dx.doi.org/10.5772/65188

cross-sectional, we cannot conclude that there is causality between social networking media, namely Facebook, and social anxiety. Does social media (time spent, pathological use, number of friends, activities) contribute to social anxiety symptomatology or do socially anxious individuals tend to use more social media? According to what we currently know, no directionality can be assumed. However, much like in the meta-analysis on Facebook and loneliness [21], in this case, social anxiety can predict social media use rather than the other way around.

Future experimental studies manipulating independent variables would be useful in order to draw more conclusions on the relationship between social networking and social anxiety. Longitudinal studies would allow us to test bidirectional relationships between the two variables and investigate time-related patterns in social networking use and social anxiety. Up to this moment, it remains unclear whether social anxiety leads to more online social networking use, or whether the relationship is opposite, with more use of online social networking predicting social anxiety. Bidirectionality between the variables could also be an important aspect to investigate, as there is preliminary evidence for both directions.

The findings of the current review have important clinical implications regarding social networking use for individuals with social anxiety. The benefits/advantages of online social networking sites should be used and investigated in order to overcome existing barriers in clinical practice. Testing interventions delivered over social networking sites in randomized controlled trials would have great implications on the development of evidence-based remote interventions.

Acknowledgements

Funding for this study was provided by a grant from the Romanian Executive Unit for Financing Education Higher Research, Development and Innovation (the "Effectiveness of an empirically based web platform for anxiety in youths", grant number PN-II-PT-PCCA-2011-3.1-1500, 81/2012) awarded to Dr. Anca Dobrean.

Author details

Anca Dobrean[1*] and Costina-Ruxandra Păsărelu[2]

*Address all correspondence to: anca.dobrean@ubbcluj.ro

1 Department of Clinical Psychology and Psychotherapy, Babeş-Bolyai University, Cluj-Napoca, Romania

2 Doctoral School Evidence-Based Assessment and Psychological Interventions, Babeş-Bolyai University, Cluj-Napoca, Romania

References

[1] Facebook Reports Second Quarter. 2016. Available from: https://investor.Facebook.com/investor-news/press-release-details/2016/Facebook-Reports-Second-Quarter-2016-Results/default.aspx [Accessed 4 August 2016].

[2] Kosinski M, Matz SC, Gosling SD, et al. Facebook as a research tool for the social sciences: opportunities, challenges, ethical considerations, and practical guidelines. Am Psychol 2015; 70: 543–556.

[3] Ryan GS. Online social networks for patient involvement and recruitment in clinical research. Nurse Res 2013; 21: 35–39.

[4] Merchant G, Weibel N, Patrick K, et al. Click 'like' to change your behavior: a mixed methods study of college students' exposure to and engagement with Facebook content designed for weight loss. J Med Internet Res 2014; 16: e158.

[5] Ruotsalainen H, Kyngäs H, Tammelin T, et al. Effectiveness of Facebook-delivered lifestyle counselling and physical activity self-monitoring on physical activity and body mass index in overweight and obese adolescents: a randomized controlled trial. Nurs Res Pract 2015; 2015: 159205.

[6] Torous J, Keshavan M. The role of social media in schizophrenia: evaluating risks, benefits, and potential. Curr Opin Psychiatry 2016; 29: 190–195.

[7] Naslund JA, Aschbrenner KA, Marsch LA, et al. The future of mental health care: peer-to-peer support and social media. Epidemiol Psychiatr Sci 2016; 25: 113–122.

[8] Hui A, Wong PW-C, Fu K-W. Evaluation of an online campaign for promoting help-seeking attitudes for depression using a Facebook advertisement: an online randomized controlled experiment. JMIR Ment Health 2015; 2: e5.

[9] Ellison NB, Steinfield C, Lampe C. The benefits of Facebook 'Friends:' social capital and college students' use of online social network sites. J Comput-Mediat Comm 2007; 12: 1143–1168.

[10] Grajales FJ, Sheps S, Ho K, et al. Social media: a review and tutorial of applications in medicine and health care. J Med Internet Res 2014; 16: e13.

[11] Laranjo L, Arguel A, Neves AL, et al. The influence of social networking sites on health behavior change: a systematic review and meta-analysis. J Am Med Inform Assoc 2015; 22: 243–256

[12] Antheunis ML, Schouten AP, Krahmer E. The role of social networking sites in early adolescents' social lives. J Early Adolesc 2016; 36: 348–371.

[13] Myhre JW, Mehl MR, Glisky EL. Cognitive benefits of online social networking for healthy older adults. J Gerontol B Psychol Sci Soc Sci. Epub ahead of print 16 March 2016. DOI: 10.1093/geronb/gbw025.

[14] Lenhart A. Teens, social media & technology overview 2015. Pew Research Center: Internet, Science & Tech. 2015. Available from: http://www.pewinternet.org/2015/04/09/teens-social-media-technology-2015 [Accessed 4 August 2016].

[15] Sampasa-Kanyinga H, Lewis RF. Frequent use of social networking sites is associated with poor psychological functioning among children and adolescents. Cyberpsychol Behav Soc Netw 2015; 18: 380–385.

[16] Sampasa-Kanyinga H, Hamilton HA. Social networking sites and mental health problems in adolescents: the mediating role of cyberbullying victimization. Eur Psychiatry 2015; 30: 1021–1027.

[17] Jelenchick LA, Eickhoff JC, Moreno MA. 'Facebook depression?' Social networking site use and depression in older adolescents. J Adolesc Health 2013; 52: 128–130.

[18] O'Keeffe GS, Clarke-Pearson K, Mulligan DA, et al. Clinical report – The impact of social media on children, adolescents, and families. Pediatrics 2011; 127: 800–804.

[19] Frison E, Eggermont S. 'Harder, Better, Faster, Stronger': negative comparison on Facebook and adolescents' life satisfaction are reciprocally related. Cyberpsychol Behav Soc Netw 2016; 19: 158–164.

[20] Marino C, Vieno A, Pastore M, et al. Modeling the contribution of personality, social identity and social norms to problematic Facebook use in adolescents. Addict Behav 2016; 63: 51–56.

[21] Song H, Zmyslinski-Seelig A, Kim J, et al. Does Facebook make you lonely? A meta analysis. Comput Human Behav 2014; 36: 446–452.

[22] Rosen LD, Whaling K, Rab S, et al. Is Facebook creating 'iDisorders'? The link between clinical symptoms of psychiatric disorders and technology use, attitudes and anxiety. Comput Human Behav 2013; 29: 1243–1254.

[23] Lin LY, Sidani JE, Shensa A, et al. Association between social media use and depression among U.S. young adults. Depress Anxiety 2016; 33: 323–331.

[24] Soon-Li Lee, Miriam Sang-Ah Park, Cai-Lian Tam. The relationship between Facebook attachment and obsessive-compulsive disorder severity. Cyberpsychology 2015; 9: 39–47.

[25] Andreassen CS. Online social network site addiction: a comprehensive review. Curr Addict Rep 2015; 2: 175–184.

[26] Hormes JM, Kearns B, Timko CA. Craving Facebook? Behavioral addiction to online social networking and its association with emotion regulation deficits. Addiction 2014; 109: 2079–2088.

[27] Koc M, Gulyagci S. Facebook addiction among Turkish college students: the role of psychological health, demographic, and usage characteristics. Cyberpsychol Behav Soc Netw 2013; 16: 279–284.

[28] Desjarlais M, Willoughby T. A longitudinal study of the relation between adolescent boys and girls' computer use with friends and friendship quality: support for the social compensation or the rich-get-richer hypothesis? Comput Human Behav 2010; 26: 896–905.

[29] Fernandez KC, Levinson CA, Rodebaugh TL. Profiling predicting social anxiety from Facebook profiles. Soc Psychol Personal Sci 2012; 3: 706–713.

[30] Rauch SM, Strobel C, Bella M, et al. Face to face versus Facebook: does exposure to social networking web sites augment or attenuate physiological arousal among the socially anxious? Cyberpsychol Behav Soc Netw 2014; 17: 187–190.

[31] Moher D, Liberati A, Tetzlaff J, et al. Preferred reporting items for systematic reviews and meta-analyses: the PRISMA statement. Ann Intern Med 2009; 151: 264–269, W64.

[32] Bodroža B, Jovanović T. Validation of the new scale for measuring behaviors of Facebook users: psycho-social aspects of Facebook use (PSAFU). Computers in Human Behavior 2016; 54: 425–435.

[33] große Deters F, Mehl MR, Eid M. Social responses to Facebook status updates: the role of extraversion and social anxiety. Comput Human Behav 2016; 61: 1–13.

[34] Landoll RR, La Greca AM, Lai BS. Aversive Peer Experiences on Social Networking Sites: Development of the Social Networking-Peer Experiences Questionnaire (SN-PEQ). *J Res Adolesc* 2013; 23: 695–705.

[35] Lee-Won RJ, Herzog L, Park SG. Hooked on Facebook: The role of social anxiety and need for social assurance in problematic use of Facebook. Cyberpsychol Behav Soc Netw 2015; 18: 567–574.

[36] Shaw AM, Timpano KR, Tran TB, et al. Correlates of Facebook usage patterns: the relationship between passive Facebook use, social anxiety symptoms, and brooding. Comput Human Behav 2015; 48: 575–580.

[37] Weidman AC, Levinson CA. I'm still socially anxious online: offline relationship impairment characterizing social anxiety manifests and is accurately perceived in online social networking profiles. Comput Human Behav 2015; 49: 12–19.

[38] McCord B, Rodebaugh TL, Levinson CA. Facebook: social uses and anxiety. Computers in Human Behavior 2014; 34: 23–27.

[39] Clayton RB, Osborne RE, Miller BK, et al. Loneliness, anxiousness, and substance use as predictors of Facebook use. Comput Human Behav 2013; 29: 687–693.

[40] Hong J-C, Hwang M-Y, Hsu C-H, et al. Belief in dangerous virtual communities as a predictor of continuance intention mediated by general and online social anxiety: the Facebook perspective. Comput Human Behav 2015; 48: 663–670.

[41] Indian M, Grieve R. When Facebook is easier than face-to-face: social support derived from Facebook in socially anxious individuals. Pers Individ Dif 2014; 59: 102–106.

http://dx.doi.org/10.5772/65188

[42] Ramirez A, Sumner EM, Hayes J. Reconnect on Facebook: The Role of Information Seeking Behavior and Individual- and Relationship-Level Factors. *Cyberpsychology, Behavior, and Social Networking* 2016; 19: 494–501.

[43] Green T, Wilhelmsen T, Wilmots E, et al. Social anxiety, attributes of online communication and self-disclosure across private and public Facebook communication. Comput Human Behav 2016; 58: 206–213.

[44] Liu C, Ang RP, Lwin MO. Cognitive, personality, and social factors associated with adolescents' online personal information disclosure. J Adolesc 2013; 36: 629–638.

[45] Aladwani AM, Almarzouq M. Understanding compulsive social media use: the premise of complementing self-conceptions mismatch with technology. Comput Human Behav 2016; 60: 575–581.

[46] Casale S, Fioravanti G. Satisfying needs through social networking sites: a pathway towards problematic Internet use for socially anxious people? Addict Behav Rep 2015; 1: 34–39.

[47] Lee EB. Too much information heavy smartphone and Facebook utilization by African American young adults. J Black Stud 2015; 46: 44–61.

[48] Rizvi AZ. Personality, social anxiety and excessive use of Facebook. Int J Psychol Behav Sci 2016; 6: 119–127.

[49] Davidson T, Farquhar LK. Correlates of social anxiety, religion, and Facebook. J Media Religion 2014; 13: 208–225.